ENERGY AND
THE ENVIRONMENT

Coal and Oil

**ENERGY AND
THE ENVIRONMENT**

Coal and Oil

JOHN TABAK, Ph.D.

☑ Facts On File

An imprint of Infobase Publishing

For Richard DiNatale, from whom I learned enough to last me a lifetime—at least so far

COAL AND OIL

Copyright © 2009 by John Tabak, Ph.D.

Facts On File, Inc.
An imprint of Infobase Publishing
132 West 31st Street
New York NY 10001

Library of Congress Cataloging-in-Publication Data

Tabak, John.
  Coal and oil / John Tabak.
    p. cm. — (Energy and the environment)
  Includes bibliographical references and index.
  ISBN-13: 978-0-8160-7083-1 (acid-free paper)
  ISBN-10: 0-8160-7083-0 (acid-free paper)
  1. Coal. 2. Petroleum as fuel. 3. Fossil fuels. I. Title.
  TP319.T33 2009
  333.8′2—dc22   2008024343

Facts On File books are available at special discounts when purchased in bulk quantities for businesses, associations, institutions, or sales promotions. Please call our Special Sales Department in New York at (212) 967-8800 or (800) 322-8755.

You can find Facts On File on the World Wide Web at http://www.factsonfile.com

Text design by Erik Lindstrom
Illustrations by Accurate Art
Photo research by Elizabeth H. Oakes

Printed in the United States of America

Bang Hermitage 10 9 8 7 6 5 4 3 2

This book is printed on acid-free paper.

Contents

Preface

Nations around the world already require staggering amounts of energy for use in the transportation, manufacturing, heating and cooling, and electricity sectors, and energy requirements continue to increase as more people adopt more energy-intensive lifestyles. Meeting this ever-growing demand in a way that minimizes environmental disruption is one of the central problems of the 21st century. Proposed solutions are complex and fraught with unintended consequences.

The six-volume Energy and the Environment set is intended to provide an accessible and comprehensive examination of the history, technology, economics, science, and environmental and social implications, including issues of environmental justice, associated with the acquisition of energy and the production of power. Each volume describes one or more sources of energy and the technology needed to convert it to useful working energy. Considerable emphasis is placed on the science on which the technology is based, the limitations of each technology, the environmental implications of its use, questions of availability and cost, and the way that government policies and energy markets interact. All of these issues are essential to understanding energy. Each volume also includes an interview with a prominent person in the field addressed. Interview topics range from the scientific to the highly personal, and reveal additional and sometimes surprising facts and perspectives.

Nuclear Energy discusses the physics and technology of energy production, reactor design, nuclear safety, the relationship between commercial nuclear power and nuclear proliferation, and attempts by the United States to resolve the problem of nuclear waste disposal. It concludes by contrasting the nuclear policies of Germany, the United States, and France. Harold Denton, former director of the Office of Nuclear Reactor Regulation at the U.S. Nuclear Regulatory Commission, is interviewed about the commercial nuclear industry in the United States.

Biofuels describes the main fuels and the methods by which they are produced as well as their uses in the transportation and electricity-production sectors. It also describes the implications of large-scale biofuel use on the environment and on the economy with special consideration given to its effects on the price of food. The small-scale use of biofuels—for example, biofuel use as a form of recycling—are described in some detail, and the volume concludes with a discussion of some of the effects that government policies have had on the development of biofuel markets. This volume contains an interview with economist Dr. Amani Elobeid, a widely respected expert on ethanol, food security, trade policy, and the international sugar markets. She shares her thoughts on ethanol markets and their effects on the price of food.

Coal and Oil describes the history of these sources of energy. The technology of coal and oil—that is, the mining of coal and the drilling for oil as well as the processing of coal and the refining of oil—are discussed in detail, as are the methods by which these primary energy sources are converted into useful working energy. Special attention is given to the environmental effects, both local and global, associated with their use and the relationships that have developed between governments and industries in the coal and oil sectors. The volume contains an interview with Charlene Marshall, member of the West Virginia House of Delegates and vice chair of the Select Committee on Mine Safety, about some of the personal costs of the nation's dependence on coal.

Natural Gas and Hydrogen describes the technology and scale of the infrastructure that have evolved to produce, transport, and consume natural gas. It emphasizes the business of natural gas production and the energy futures markets that have evolved as vehicles for both speculation and risk management. Hydrogen, a fuel that continues to attract a great deal of attention and research, is also described. The book focuses on possible advantages to the adoption of hydrogen as well as the barriers that have so far prevented large-scale fuel-switching. This volume contains an interview with Dr. Ray Boswell of the U.S. Department of Energy's National Energy Technology Laboratory about his work in identifying and characterizing methane hydrate reserves, certainly one of the most promising fields of energy research today.

Wind and Water describes conventional hydropower, now-conventional wind power, and newer technologies (with less certain futures) that are being introduced to harness the power of ocean currents, ocean waves, and the temperature difference between the upper and lower layers of the ocean. The strengths and limitations of each technology are discussed at some length, as are mathematical models that describe the maximum amount of energy that can be harnessed by such devices. This volume contains an interview with Dr. Stan Bull, former associate director for science and technology at the National Renewable Energy Laboratory, in which he shares his views about how scientific research is (or should be) managed, nurtured, and evaluated.

Solar and Geothermal Energy describes two of the least objectionable means by which electricity is generated today. In addition to describing the nature of solar and geothermal energy and the processes by which these sources of energy can be harnessed, it details how they are used in practice to supply electricity to the power markets. In particular, the reader is introduced to the difference between base load and peak power and some of the practical differences between harnessing an intermittent energy source (solar) and a source that can work virtually continuously (geothermal). Each section also contains a discussion of some of the ways that governmental policies have been used to encourage the growth of these sectors of the energy markets. The interview in this volume is with John Farison, director of Process Engineering for Calpine Corporation at the Geysers Geothermal Field, one of the world's largest and most productive geothermal facilities, about some of the challenges of running and maintaining output at the facility.

Energy and the Environment is an accessible and comprehensive introduction to the science, economics, technology, and environmental and societal consequences of large-scale energy production and consumption. Photographs, graphs, and line art accompany each text. While each volume stands alone, the set can also be used as a reference work in a multidisciplinary science curriculum.

Acknowledgments

The author is particularly grateful for the help of the following individuals: Charlene Marshall, for generously sharing her story; Ellen Smith, owner and editor of *Mine Safety and Health News,* for her insights and wisdom; Judy Bonds, of *Coal River Mountain Watch,* for sharing with me her insights into the environmental and social costs of the coal economy; Gail Christian, for her patience and support; Elizabeth Oakes, for the fine photo research; and Frank K. Darmstadt, executive editor, for the vote of confidence.

Introduction

Energy is one of the central issues of the 21st century, and coal and oil are the world's two most important sources of primary energy. *Coal and Oil* describes humanity's complex relationships with these fuels: the characteristics of these energy sources, the ways in which they are used, and the technical, social, policy, and environmental consequences of large-scale coal and oil consumption.

A multitrillion dollar infrastructure has been created to locate, produce, transport, process, and burn coal and oil. This infrastructure has made modern life possible. In some of the world's largest economies, coal-fired power plants generate at least half of all electricity, and in almost all nations, transportation is synonymous with oil consumption. To understand why the world is so dependent on coal and oil, one must understand what is good about these fuels, and that is a major focus of this book.

Despite the benefits associated with the large-scale consumption of coal and oil—and there are many and they are substantial—from an environmental point of view, humanity's dependence on these fuels has been a disaster. In addition to environmental effects, significant social costs associated with the production and consumption of coal and oil have been imposed on the politically less powerful—even as the benefits associated with access to abundant and affordable supplies of fossil-fuel energy are shared by many. Identifying and describing these environmental and social costs is the second major focus of this book. (See the interesting and highly personal interview with Charlene Marshall, a member of the West Virginia House of Delegates and Vice Chair of the Select Committee on Mine Safety, about the human costs of coal.)

Chapter 1 deals with the history of coal, a history of rapid technological progress and brutal and exploitive working conditions. Chapter 2 describes some of the physical and chemical characteristics of coal, and concludes with an introduction to the geology of this important mineral. To appreciate how central coal is to modern life, one must understand the astonishing scale at which it is consumed as well as the associated environmental and social costs. That is the goal of chapter 3. The chemistry of coal combustion and some of its environmental consequences are described in chapter 4, and the technology used to convert coal into electricity is described in chapter 5. Because coal is so abundant, many nations are seeking new ways to use it. Describing some of the most promising of these technologies is the goal of chapter 6, and chapter 7 describes some relationships between coal consumption and national energy policies.

The second half of *Coal and Oil* is devoted to oil, and its structure mirrors that of the first half. Chapter 8 is devoted to the history of oil and chapter 9 to the geology and chemistry of crude oil. Chapter 10 describes how oil is produced and transported, and chapter 11 describes the role of refineries and the chemistry of combustion.

Patterns of oil consumption and the environmental costs associated with the oil economy are described in chapters 12 and 13. Chapter 14 discusses government policies, oil markets, and the price of petroleum.

The almost abject dependence of oil producing and consuming nations on the oil trade, and the central role of coal in the manufacture of electric power, raise difficult questions about the environment, technology, and the future of the world economy. These questions have no easy answers, but as supplies of oil dwindle and climate change accelerates, the search for answers will become increasingly urgent.

PART I

Coal

A Brief History of Coal

Humanity has long recognized the value of coal. Many people have worked, suffered, and even died to obtain it. Many of the machines and processes that formerly depended upon coal as an energy source are now obsolete, but coal is as necessary today as it has ever been. This chapter examines some of the ways that the search for coal and the switch to coal have changed history.

But coal is more than fuel. Coal production has also had very important social consequences. Coal miners formed some of the earliest labor unions as they sought to procure better pay and safer working conditions. These struggles continue today. This chapter also describes some of the social history of this essential energy source.

THE EARLY HISTORY OF COAL

Coal has been used in many societies throughout the world for a very long time. Upon his return from China, Marco Polo

Miners at the entrance to the Blue Canyon coal mine, which operated from 1891 to 1918 in Washington State *(Whatcom Museum of History and Art)*

(1254–1324) described a black stone that burned like wood. This material, evidently unfamiliar to him, had been used by the Chinese—perhaps as early as 1000 B.C.E.—to smelt metal. Romans were mining coal in England by 400 C.E., and long before the arrival of Columbus, Native Americans of the Southwest were using coal. In most early societies, the alternative to coal was wood. Historically, of course, wood has also been an important source of energy, but there were two important reasons that some early societies turned to coal.

First, when measured by weight, coal releases more heat when burned than the same amount of wood. It is difficult to be precise about the amount of heat released by burning coal because coal's energy content—also called its *heating value*—depends on how much carbon is in the coal, its moisture content, and the amount of noncombustible solids found in the coal. All of these characteristics vary widely from one coal deposit to the next.

Second, because wood has a lower heating value than coal, early energy consumers who depended on wood for heat had to burn a lot of it. In addition to its value as a fuel source, wood was also used to build houses and ships and many smaller, more delicate objects such as furniture and musical instruments. (By contrast, coal's only value was as a fuel.) Consequently, technologically simple societies often exhausted their supplies of wood. By 1300 c.e., for example, many English forests had been consumed as demand for wood caused the populace to denude the landscape. That was the reason that the English turned to coal so early in their history. They needed an alternative fuel source, and they had large coal deposits, many of which were located at or near the surface. The shift from wood to coal in Britain would eventually have profound consequences for the world.

To understand some of the consequences of the British reliance upon coal, it helps to know a little about the geology of coal deposits. Earth's surface consists of layers. These layers have different chemical and physical properties. As a general rule, the layers are formed horizontally, but forces within the Earth often bend and fold these layers, sometimes contorting them into complicated shapes. Erosion removes parts of the uppermost layers and exposes some of the lower layers to the atmosphere. When the British first began to turn to coal there were many outcroppings of it, places where erosion had exposed a layer, or seam, of coal to the open air. Anyone wanting some of this coal could chip at the exposed seam until sufficient pieces were removed for burning. Coal was literally there for the taking.

But as British demand for coal grew, the exposed seams of coal became increasingly depleted. By the 17th century, British miners began to follow the seams underground. Men, women, and children equipped with picks and baskets would chip away at the face of a coal seam for 12 or more hours per day. As they dislodged pieces of coal, chunks large and small were placed in baskets that would later be carried to the surface. These were dirty, dangerous, and debilitating jobs.

As the miners followed the coal seam deeper into the ground, they would sometimes find themselves working below the water table. As a result, water often accumulated within the mines. In order to extract these deeper coal deposits, the water had to be pumped out of the mine while the miners worked. Water removal imposed an additional and very substantial burden on the mine operators. Sometimes, when measured by weight, more water than coal was removed from a working coal mine, so that keeping the mines from flooding required at least as much effort as mining the coal.

The problem of clearing water from coal mines served as an impetus for some of the most important inventions in the history of humanity. In particular, it inspired the British engineer and inventor Thomas Savery (1650–1715) to invent the first steam-powered pump. His design was quickly improved upon by another British engineer and inventor, Thomas Newcomen (1663–1729), whose version of the steam engine powered British pumps for more than half a century. Early steam engines were huge—they weighed tons—and they were used almost exclusively as pumps to clear coal mines of water. These early pumps were also exceedingly inefficient—that is, they required large amounts of fuel to do comparatively small amounts of work. Despite its shortcomings, Newcomen's engine made it possible to more efficiently mine coal from deposits located beneath the water table.

As important as Newcomen's engine was, the fundamental change to British society occurred during the latter half of the 18th century as a result of work by the British inventor James Watt (1736–1819) to improve steam-engine technology. Watt's first efforts were aimed at modifying Newcomen's design by making it more efficient. In fact, part of the payment that Watt received for his early engines was computed in terms of the savings in coal enjoyed by the mine operator. Soon, however, Watt's engines found applications in many industries. As steam engines became more commonplace, the demand for coal soared. New inventions, such as steamboats and trains, further spurred the demand for coal.

Another important market for coal developed from the need for street illumination. Beginning in the latter years of the 18th century, the British inventor William Murdock (1754–1839), who began his professional life in the employ of James Watt, began experimenting with *coal gas*. Coal gas is produced by heating coal in a container depleted of oxygen. Coal contains substances called volatiles, materials that can be converted into gas by heating. (The coal must be heated in a low-oxygen environment to prevent the combustible volatiles from burning prematurely.) The combustible gas mixture, which was produced in facilities called coal works, was initially used almost exclusively for street lighting. Pipelines were run from the coal works to street lamps. Later, factories and homes connected to the system. Coal works, trains, boats, and metallurgical industries all demanded coal. Britain became the world's first industrialized nation, a society powered almost entirely by coal.

The manner in which coal was mined also had another effect on British society. British coal miners formed some of the first labor unions, and their unions created important changes in Britain and other nations, particularly the United States. To see why coal miners played such a pivotal role in the union movement, keep in mind that coal mining was dirty, dangerous work and that miners were initially very poorly paid. The work was so dangerous and so unhealthy that given a choice most people preferred to stay out of coal mines. This explains why, beginning in the 17th century, some mine owners and some governments colluded to force many men, women, and children to spend their lives toiling in the mines. The demand for coal was huge, but the voluntary workforce was small.

The treatment endured by the Scots was especially severe. In 1606, the Scottish Parliament passed a law that criminalized those miners that tried to quit the mines. The offense was described as theft "of themselves." The goal of the 1606 law was to enslave the miners. A miner that sought to escape could, if caught within a year and a day from the date of escape, be severely punished, and anyone hiring an escaped miner within this time frame faced a £100 fine. The one-year-and-a-day restriction was later removed, making any escaped miner a criminal for life. Even these punishments were not enough to maintain the workforce. In order to meet production quotas, the Scottish parliament soon passed a law that enabled mine owners to seize those identified as vagabonds—men, women, and their children—and force them to work in the mines. Those who were caught faced a life sentence in the mines. Orphans often suffered a similar fate.

In England the situation was less dire: Miners were required to sign year-to-year contracts that bound them to work in a mine and promised severe penalties for all that tried to leave early. In all cases, the work week was six days long and lasted at least 12 hours a day. Miners rarely saw the Sun.

As the miners tunneled deeper, methane, an invisible, odorless, and highly flammable gas, claimed the lives of ever more miners, who, given the technology of the time, had no option but to work by candlelight. The working conditions were shocking—even to many observers of the time. Labor shortages persisted. This was the dirtiest, most dangerous job in Scotland, and those who began work in the mines surrendered their freedom and that of their families, often for the remainder of all of their lives. Volunteers were, not surprisingly, hard to find.

In 1775, the British parliament passed an act that is sometimes called the Miner's Emancipation Act. The following is a brief excerpt from the first paragraph ("collier" is a synonym for coal miner and a "colliery" is a coalmine):

> . . . be it enacted that after the first day of July [1775] . . . no person, who shall begin to work as a collier, coal-bearer, or salter, or in any other way in a colliery or salt-work, in Scotland, shall be bound to such colliery or salt work, or to the owner thereof, in any way or manner different from what is permitted by the law of Scotland with regard to servants and laborers; and that they shall be deemed free . . .

Evidently, emancipation did not apply directly to the miners already working in the mines but only to those about to begin work. As indicated in the excerpt, only those beginning work in the mines after July 1, 1775, were to be considered free. Later paragraphs within the act stated that those already in a state of slavery within the mines were to be freed gradually. (The owners worried that if all of the miners were freed simultaneously, they would leave *en masse* and there would not be enough workers to maintain production.)

According to the act, colliers under the age of 21 on July 1, 1775, for example, were bound to their colliery for an additional seven years. And in order to obtain their freedom, they first had to obtain a decree of the sheriff court, an official act that permitted the mine owner to object. The goal of the 1775 act was not to free miners but to make the work in the Scottish mines more attractive in the hope that more individuals would willingly volunteer to labor within the mines. The change in the law was the result of activism on the part of the coalmine owners not coalmine workers. The miners, in fact, had little political power and their efforts at protest were sporadic and often ineffectual. The 1775 emancipation act—officially called "An act for altering, explaining, and amending several acts of the parliament of Scotland, respecting colliers, coal-bearers, and salters"—had little immediate effect on the labor shortage. The working conditions within the mines were so terrible that it would be decades before the shortages began to ease. Even after 1775, Scottish coal mining remained an industry built on slavery and brutality, and an act that would free all Scottish miners unconditionally was not passed until 1799.

English miners found another and more immediate way to better protect their interests: They organized so that they could deal directly with the mine owners. The first English miners' strike occurred in 1765 as workers successfully protested attempts by owners to increase the length of the one-year contract under which they had long worked. As the organizing sophistication of the miners increased, they formed larger and more powerful associations. But progress in improving working conditions remained slow. A British official, acting to enforce an 1833 act of Parliament called the Factory Act, visited a coal mine, and comparing the conditions of the miners with the grim conditions of factory workers, wrote:

> . . . the hardest labor in the worst room, in the worst conducted factory, is less hard, less cruel, less demoralizing than in that of the best coal mines.

In 1842, British miners formed the Miners' Association of Great Britain and Ireland, one of the first true trade unions in history. British miners with union experience soon made their way to the United States, bringing technical and organizational skills to American mines.

THE COAL INDUSTRY IN THE UNITED STATES

During colonial times and during the first decades after the Revolutionary War, wood was the fuel of choice in the United States. The population of the United States was small, and the forests must have seemed inexhaustible as a source of fuel. But while wood and charcoal, which is derived from wood, were the principal sources of heat energy early in the nation's history, there were some who attempted to use coal right from the start.

By 1740, there were several coal mines in operation along the upper Potomac River near the border of Maryland and Virginia—these were worked by slaves—and coal had been discovered in Pennsylvania by 1750. Early attempts to mine coal were primitive even by the standards of the time. The expertise that had already been developed in Britain had not yet made its way across the Atlantic. In fact, during colonial times, most of the coal burned in cities along the eastern seaboard was imported from Britain. But the domestic market for coal grew steadily: Blacksmiths required coal, and early metal manufacturers recognized the potential of coal as a source of heat. During the first few decades of the 19th century, cities began to build coal works to supply gas for street illumination, and trains and steamboats, which had originally run on wood, began to burn coal. The domestic mining industry grew rapidly to meet demand.

To understand the conditions under which the United States mining industry developed, it helps to more closely examine the conditions under which miners worked. Most early mines in the United States began at an outcropping of coal and tunneled down beneath the surface. When mining underground, a breathable atmosphere is difficult to maintain because tunnels are difficult to ventilate. Moreover, coal mining creates dust. Over time, the dust, which the miners had no choice but to inhale, destroyed their lungs and their lives. In addition, toxic gases of various types are sometimes released as coal is mined. Some of these gases are also explosive, and because the miners worked by candlelight, explosions were fairly common.

Early mines depended only on natural ventilation to remove the gases that accumulated within the mine and to furnish fresh, breathable air for the miners. Natural ventilation, which was inefficient under the best of circumstances, occurred because the air in deeper mines tended to be warmer than air at the mine's entrance. Because warm air tends to rise, the air inside the mine would slowly flow to the surface as cooler surface air sank into the mine. But this effect depends on the existence of cooler surface air. During the warmer months, warm surface air often overlays warm air within the mine, and there was little natural ventilation. The toll that bad ventilation, explosions, and *roof falls* took on the early mining workforce was terrible.

Despite the difficulties, coal production increased steadily throughout the 19th century and into the first few decades of the 20th century, as new markets were created for this extremely valuable source of energy. By the end of the 19th century almost every

⏻ Life and Death in Early Coal Mines

The 19th-century American public was occasionally made aware of the terrible price paid by miners while extracting the coal upon which so much depended. In 1869, at the Avondale mine in Plymouth, Pennsylvania, a fire asphyxiated 110 miners. In 1883, a sudden rush of surface water into a mine near Braidwood, Illinois, drowned 69 miners. In 1907, in Monongah, West Virginia, a mine explosion killed 362 miners. These tragedies and several others were widely reported in the newspapers, but they hardly conveyed the scale of the disaster occurring within the nation's mines. Between the years 1839 and 1914 more than 61,000 miners were killed at work. The leading cause of death was roof falls, which occur when part of the mine collapses on the workers inside. Roof falls generally kill one or perhaps a few miners at a time, and so the great majority of mining deaths occurred out of sight and out of mind of all but the miners and their families. In addition to the deaths that occurred within the coal mines, many miners perished from what was once called "miner's asthma," a general term that was used to describe the respiratory problems brought about by long-term exposure to the bad air within the mines.

But if the plight of coal miners received little notice in the press, miners and others worked hard to improve mine safety. For a long time their success was limited. In fact, for many years innovations in safety had an overall effect that was the opposite to what had been intended. The problem was that each innovation enabled the miners to dig deeper or to work in areas that would have been impossible before the safety innovation had been introduced. Two examples illustrate how safety devices, even when they worked as intended, often failed to improve safety.

large city and most small U.S. cities had built coal works to supply coal gas. Coal was used in metallurgy from 1840, and by the end of the 19th century the iron and steel industry had become one of the biggest industries in the nation, an industry that required enormous

First, to reduce the hazard of mine explosions, miners switched to the Davy safety lamp, designed by the British chemist Sir Humphrey Davy (1778–1829) in 1815, to illuminate their workplace. The safety lamp reduced the possibility that a lamp's flame would ignite any methane present in the mine. (Methane, an explosive gas, is often present in coal mines.) Davy's lamp worked as designed; it reduced the risk that a miner's lamp would ignite the methane. But armed with the safety lamp, miners worked in places that they would have avoided had they not had access to the lamp. As a consequence, even though the lamp reduced the pos-sibility of a methane explosion in any given situation, the total number of mine explosions did not diminish, because the miners' exposure to environments with dangerously high levels of methane increased.

Second, miners began to ventilate the mines more effectively. Furnaces came first. These were placed at the bottom of a shaft. A separate intake shaft was sunk some distance away from the shaft in which the furnace was located. The furnace would heat the air causing it to rise within its shaft. This would draw fresh air down the intake shaft, and the air would circulate through the tunnels connecting the two shafts. But furnaces were fire haz-ards surrounded by fuel. The disaster at Avondale, Pennsylvania, described in the preceding paragraph, was started by a ventilation furnace.

During the 1870s, fans replaced furnaces, enabling miners to tunnel still further underground. The miners often found themselves working ahead of the ventilated zones. The mine workplace remained poorly ven-tilated. Mine safety was an important impetus to the formation of mine workers' unions, and it remains a major focus of union activity today.

quantities of coal. And during the 19th century, coal became the principal transportation fuel, used by locomotives and ships, and, of course, coal was also an important heating fuel during this time. In 1882, Thomas Edison's coal-fired electric generating station, the first of its kind, began operation. Coal production maintained pace with the ever-increasing demand. In 1905, the United States produced 351 million *short tons* (319 *metric tons*) of coal, far more than any other nation on Earth, but it was not just the amount but the rate of increase that is so impressive: Between 1900 and 1920 U.S. coal production tripled.

Throughout the 19th century, United States miners worked to form effective unions to represent their interests. Some of the more important issues included better pay, restricted work hours, a safer working environment, health benefits, and insurance. One of the earliest organizations, with a membership of approximately 5,000 miners, formed in 1849 in Pennsylvania under the leadership of the British immigrant John Bates (1809–50). Bates had apparently been a member of a British social and political organization called the Chartist movement prior to moving to the United States. (The Chartist movement had broad public support in Britain, but disbanded before it achieved its goals. Almost all of the elements in its charter were, however, eventually incorporated into British law.) The union he led in the United States was called Bates' union. The same year the union was formed, the miners entered into a bitter strike for higher wages. They were unsuccessful, and the union disbanded.

Miners continued to form local organizations. One less-controversial attempt to improve their situation was through the formation of so-called benevolent associations. Members paid dues, and some of the money was used to pay the family of a member in the event that the member died. Funds were also sometimes distributed to a member in the event that the recipient became disabled. Benevolent associations spread the risks associated with mining among the membership, but they did not reduce the risks.

The first national miners' union began in St. Clair County, Illinois, where, in 1860, miners were paid 2.5 cents per 80 pounds of coal produced. Toward the end of that year, however, all the mine operators in the region announced that they intended to cut the payment by one-fourth of a cent (to 2.25 cents) per 80 pounds of coal. The miners submitted to the pay cut and continued to work. A few weeks later, the mine operators announced a second one-fourth cent cut, and the miners struck. They were successful in rolling back the price cuts and decided to form a permanent union to protect their interests. Originally, the union was restricted to the mines located in St. Clair and Madison Counties in Illinois and nearby St. Louis County in Missouri. Soon, however, they were making contacts with local coal miners' unions farther away. To increase their bargaining power, they soon decided to form a national union called the Miners' Association, a name that was later changed to the American Miners' Association. The first meeting of the organization, described in the February 1, 1861 issue of the local *Belleville Advocate* newspaper, took place in Belleville, Illinois, on January 18, 1862. Two men with experience in British coal mines, Daniel Weaver (1824–99) and Thomas Lloyd (1824–96), emerged as leaders of the new union. As with John Bates, both Weaver and Lloyd had also apparently been members of the Chartist movement in Great Britain.

Support for the new union varied from region to region. Organizers were able to unionize some mines in Pennsylvania, but mines in the South, which used slave labor, were out of reach. Because the amount of support that the union received was uneven, collective bargaining would, under normal circumstances, have been difficult, but the union was formed under extraordinary circumstances. The Civil War (1860–65) caused a tremendous increase in the demand for coal. Many miners left their jobs and went to war, and as a consequence there was insufficient manpower to meet the demand for coal. In response to the labor shortage, both the price paid for coal and the wages that miners received soared. During the Civil War, average wages increased more than threefold. But the situation was temporary. When the war ended, demand for coal decreased, and the size of the labor force increased as veterans returned to the mines. The American Miners' Association was unable to respond in a way that met the needs of its membership, and the union collapsed entirely by 1873.

Not every regional union had been absorbed by the American Miners' Association during its period of expansion. Most notable was the Workingman's Benevolent Association, a union formed by Pennsylvanian coal miners. The Workingman's Benevolent Association responded in highly sophisticated ways to the contraction of the coal markets following the Civil War. Their leader was John Siney (1831–79), an Irish-born miner who grew up in England. As the coal market contracted, the members of this union struck and forced the local mine operators to regulate coal production in order to maintain prices and stabilize the miners' wages. This approach worked for a few years, but by 1873 the coal market had contracted 40 percent from its wartime high, and operators, who had formed their own organization, opted to slash wages. The miners struck, and maintained their strike for six months. In the end, however, they were unsuccessful, and this union collapsed as well.

Miners continued to organize in order to obtain better pay and safer working conditions. The United Mine Workers of America (UMWA), the principal union representing mine workers today, was established in 1890. The UMWA survived where its predecessors had not and became one of the largest and most powerful unions in the United States. After numerous difficult strikes, UMWA members won safer working conditions, better pay, health and retirement benefits, and a number of other benefits that many U.S. workers now take for granted. (The eight-hour work day, for example, is something for which UMWA members had to fight.) The UMWA was at the forefront of the American labor movement, and its successes had an effect throughout the broader economy, one more illustration of the very profound effects that the coal industry has had on the history of the United States.

The UMWA today represents workers in a number of occupations—healthcare workers and truck drivers, for example—but its

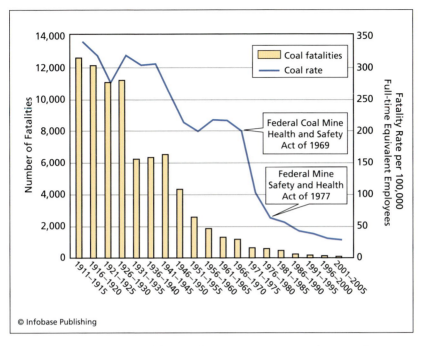

The last century was marked by steady improvement in coal mine safety.

power as a representative of miners' interests is not as great as it once was. The number of miners has plummeted as mining operations have become increasingly mechanized, and miners at many mines have opted to not join the union. Despite its difficulties, the UMWA remains an important force in the labor movement as it seeks to adapt to changing economic conditions and attitudes.

Geochemistry and Geology

There are different ways to burn coal to produce power. Different technologies have different efficiencies and affect the environment in different ways. Before one can appreciate the value of today's coal combustion technologies, one must know something about the chemical composition of coal. That is one of the goals of this chapter. There are also different ways to mine coal. Different mining technologies also have different efficiencies and different environmental impacts. In order to appreciate the differences and difficulties associated with modern coal mining, one should understand a little about the geology of coal. That is the other goal of this chapter.

THE CHEMISTRY OF COAL

Coal is not one mineral but many. The characteristics of a sample of coal depend very much on where it was mined; they depend on

This coal deposit in northeast Kazakhstan has seams of coal that are 560 feet (170 m) thick—it is mined by bucket wheel excavators. *(Peter Gunn)*

the conditions under which the coal deposit was formed. There are, not surprisingly, some characteristics that all samples of coal share: Coal is a combustible rock formed from the partially decomposed remains of plants and animals, and when measured by weight, the principal element of which coal is formed is carbon. But within this very broad description, there are many different types of coal. Of particular importance in the classification of coals are the following four properties:

Samples of coal can be classified according to their *volatile content*. As previously mentioned, volatiles are materials present in coal that can be separated by heating. Volatile content is determined by heating the coal to 221°F (105°C) to remove any water present in the sample, measuring the weight of the dry sample, and then heating the sample to 1,741°F (950°C) and noting the change in weight of the sample from the first measurement to the second. Some samples of coal are composed of only 5 percent volatile matter when measured by weight, but some are composed of more than 30 percent volatile matter.

Historically, volatiles formed the basis for coal gas, described in chapter one, which was the principal gaseous fuel of the 19th and early 20th centuries. Today, engineers are again devising methods for extracting and processing the volatiles to manufacture a cleaner-burning gas.

The second characteristic by which samples of coal can be classified is the fixed carbon content of the coal. Coal contains volatiles, fixed carbon, and noncombustible material called *ash*. The percentage of fixed carbon in a sample of coal can vary widely from one coal deposit to the next. Some deposits of *lignite*, a type of soft brown coal, may contain only 25–30 percent fixed carbon, while *anthracite*, a type of hard black coal, may contain as much as 98 percent fixed carbon. Burning fixed carbon releases a great deal of heat as well as large amounts of carbon dioxide, a potent *greenhouse gas.*

A third characteristic of coal that is important in evaluating its potential as a fuel is the amount of sulfur present in a sample. Sulfur content varies from less than 1 percent to about 4 percent depending on the origin of the coal. Sulfur content is important because when the coal is burned the sulfur combines with oxygen to produce sulfur dioxide, written SO_2. (The letter *S* is the chemical symbol for sulfur, and the letter *O* is the chemical symbol for oxygen. The subscript 2 means that there are two oxygen atoms chemically bound to each sulfur atom in the sulfur dioxide molecule.) Sulfur dioxide is one of the causes of acid rain; it can damage buildings, and it can irritate lungs. A great deal of effort has been expended finding ways to process coal so as to reduce its sulfur content prior to combustion, and more effort has been expended devising ways to remove sulfur dioxide from the combustion gases prior to venting them to the atmosphere.

A fourth characteristic of coal is the amount of noncombustible material, called ash, that remains after the coal has been burned.

Ash is composed of a wide variety of materials. Ash content ranges from about 5 percent for coal from the Powder River Basin, which spans parts of Wyoming and Montana, to as much as 50 percent for certain deposits of coal in India. On average, coal burned in the United States to produce electric power contains anywhere from 5 to 19 percent ash with an average of 10 percent. This means that burning one short ton (907 kg) of coal produces about 200 pounds (91 kg) of ash. Ash can corrode boiler components, and it contains trace amounts of toxic heavy metals, but the main problem with ash is that there is so much of it. Currently, United States' coal-burning power plants generate about 120 million short tons (110 metric tons) of coal combustion products annually, enough to fill one million railroad cars. (Coal combustion products consist principally of ash together with boiler slag and the waste products generated by the emissions control systems employed at most power plants.) Disposing of this material in a way that minimizes its environmental impact is an important environmental problem.

Although the four characteristics of coal already listed are the ones of most importance to this volume, there are many other characteristics that deserve a brief mention. Moisture content, for example, is important in determining the value of coal. When measured by weight, the amount of water in coal can vary from less than 2 percent to more than 30 percent. Lower moisture content means a hotter flame. Of course, the coal can always be dried before it is burned, but this requires energy.

Coal also contains many trace "impurities." In fact, most of the naturally occurring elements of the periodic table have been detected in coal—with different samples containing different elements. Mercury is an example of an important trace impurity. Although the amount of mercury in a given sample of coal is always small, the amount of coal burned is so enormous that significant amounts of mercury are emitted into the atmosphere each year due solely to the

(continued on page 24)

A Coal Economy

Many writers claim that the world's energy supply is becoming increasingly "decarbonized," another way of saying that older fuels, which have higher ratios of carbon atoms to hydrogen atoms, have been deemphasized in favor of fuels with lower carbon–hydrogen ratios. The progression is supposed to begin with coal, which consists largely of carbon. Historically, the next fossil fuel to come into widespread use was petroleum. Oil has a lower ratio of carbon atoms to hydrogen atoms than coal does. Burning oil produces 180 pounds of carbon dioxide per one million Btu (82 kg/million kJ) of thermal energy. By contrast, burning coal yields roughly 200 pounds of carbon dioxide per million Btu (91 kg/million kJ) of coal. Natural gas, which came into widespread use after oil, has a lower carbon-to-hydrogen ratio than even oil. Producing one million Btu through the combustion of natural gas generates only about 120 pounds of carbon dioxide (54 kg/million kJ). Hydrogen gas, which is often touted as the fuel of the future, has no carbon at all. In this view, coal is the old-fashioned high-carbon fuel.

But the fuel supply has not been decarbonized. In many countries, most notably the United States and China, two of the world's biggest energy users, coal is still an extremely important fossil fuel. In the United States, for example, coal consumption has climbed fairly steadily over the last century even as many once important coal-based industries have died out or switched to other fuels. Currently, in the United States about 20 pounds (9 kg) of coal are consumed per person per day. It is easy to see why: Coal is too inexpensive to ignore. While the exact numbers vary from day to day, purchasing enough coal to generate 1 million Btus (1.054 billion joule), costs only a small fraction of a million Btus' worth of natural gas or gasoline, for example, and the price disparity between coal and other fossil fuels continues to grow.

Today, more than 90 percent of all coal burned in the United States is consumed by the electric-power industry, and coal-fired power plants now produce about 50 percent of the nation's electricity. In 1990 domestic coal production exceeded one billion short tons (907 million metric tons) per year for the first time, a rate of production that has since become ordinary.

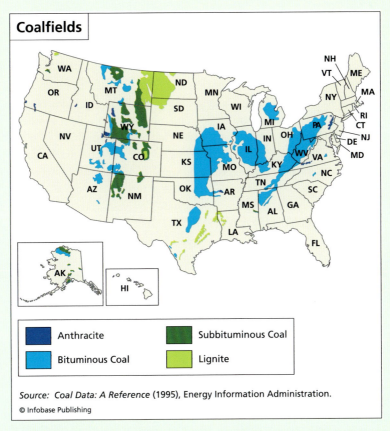

Coalfields

■ Anthracite	■ Subbituminous Coal
■ Bituminous Coal	■ Lignite

Source: Coal Data: A Reference (1995), Energy Information Administration.

© Infobase Publishing

U.S. coal reserves. Coal lies beneath approximately 13 percent of the nation's landmass. *(Modeled after EIA)*

Enormous sums of money have already been spent building coal-fired power plants in developed and developing nations alike. Many of these plants will continue to produce electricity for decades, guaranteeing that

(continues)

(continued)

coal will remain in demand for the foreseeable future. Additional large sums of money are being spent attempting to find ways to use coal to produce gaseous and liquid fuels. The technology already exists to gasify coal. Plants exist to produce methane (natural gas) from coal, and a synthetic gas that consists of carbon monoxide and hydrogen gas for power generation, and it is technically possible to produce large quantities of pure hydrogen gas from coal. The technology also exists to obtain synthetic petroleum from coal, which can be used in the transportation sector. If these technologies are implemented on a large scale—and this becomes more likely as oil and natural gas become ever more expensive—coal use would soar to much higher levels that those of today.

The reserves of coal in the United States are vast. The United States could, for example, easily continue to meet its current demand for coal for centuries using only its own reserves. These statements are also true of China and several other large energy-hungry countries. Coal was the first and it may also be the last abundant and easily obtainable fossil fuel. There is every indication that in the future the fuel supply will become increasingly "carbonized" as coal becomes an even more important part of the energy mix than it is today.

(continued from page 21)

burning of coal. Mercury is highly toxic and ingestion of mercury can cause a variety of neurological disorders, especially in young children. Once emitted from a smokestack, mercury eventually settles to the ground. Today elevated levels of mercury can be detected in fish throughout the northeast section of the United States. Much of this mercury was emitted by coal-burning power plants in the United States Midwest.

Coal is an energy-rich fuel that is laced with materials that are potentially harmful. Burning coal releases some of these materials into the broader environment; it also creates other materials—greenhouse gases, for example—that were not present in the sample prior to the time the coal was burned. Despite these drawbacks, coal is currently being burned at a rate higher than at any time in history, and that rate is increasing. The reason is geological. Located within the borders of some of the world's most energy-hungry countries are enormous deposits of coal.

THE GEOLOGY OF COAL

A lot has been written about the process by which coal is formed, but there is a lot about the process that remains unknown. What is known is that today's coalfields are composed of the remains of plants and animals that lived millions of years ago. The proof is that fossils of some of these organisms are common and easily visible in some coal deposits. The types of fossils present in the coal are also helpful in determining the environment in which the deposit formed and the period in which it formed.

Coal originates in swamps. The process of coal formation begins when some of the remains of the plants and animals that live in the swamps become submerged and partially decomposed. Eventually, these remains are covered by the remains of other creatures, and slowly the layers accumulate. Coal beds reveal a great deal about the history of the ancient swamps in which they were formed. Some coal beds, for example, reveal that there were occasional droughts. The vegetation became dry, and the swamps burned before they were again inundated with water. The proof is that layers of charcoal, the remains of ancient fires, are present in the coal deposits.

Sometimes a swamp would become flooded. Water would rush over the swamp, leaving a layer of clay to settle across the top of the partially decomposed plant and animal matter that had accumulated on the floor of the swamp. Clay cannot be burned and it does not decay. Consequently, the clay that was deposited millions of years ago is still present in the coal deposit. Its presence contributes

The fossils found in coal provide information about when the deposit was formed and the conditions that prevailed when it was created. *(University of Birmingham, School of Geography, Earth & Environmental Sciences)*

to the ash generated at power stations when the coal is burned to produce electricity.

Some swamps formed near the ocean. They were regularly inundated with seawater. The ocean water contained sulfur compounds. Some of the difference between low-sulfur coals and high-sulfur coals is thought to be determined by whether the swamp water was salty or fresh.

Eventually, the swamps were buried, sometimes beneath thousands of feet of sediment. Meanwhile, water flowed through the deposits. The water deposited still other chemicals, permanently changing the chemical composition of the coal that would result from this process. As the remains of the already ancient swamp sank further beneath the surface, the plant and animal remains were exposed to ever increasing temperatures and pressures. Over time, the partially decomposed remains of these ancient creatures become coal. This process is called *coalification*.

Because coal deposits are formed at different times and subjected to different degrees of heat and pressure, the coalification process proceeds at different rates for different deposits. The degree of coalification is used to classify coals. The least transformed deposits—that is, the deposits that are most similar to the peat from which they formed—are called lignite, also known as brown coal. As the rank increases, so does the percentage of fixed carbon. In order of increasing rank, there is *subbituminous, bituminous,* and anthracite. Higher ranks of coal are generally harder and shinier than lower ranks and have somewhat higher heating values.

The changes caused by coalification are not just chemical. The pressure and heat to which the coal deposits are subjected compact the deposits, thereby increasing their density. It is a commonly quoted estimate that from three to 10 feet (1–3 m) of layered material are required to produce a one-foot- (30-cm-) thick deposit of bituminous coal. Some coal beds are less than one foot (30 cm) thick—these are generally uneconomical to mine—and some exceed 100 feet (30 m) in thickness. The deposits in the coal mining regions of the eastern United States, for example, generally vary from three to six feet (1–2 m), and in Wyoming the average thickness of currently mined coal beds is about 60 feet (20 m).

The remains of those ancient swamps are not only compacted and chemically changed, but they are sometimes also tilted and folded. Many coal beds are formed from material deposited before the dinosaurs walked the Earth. In the interim, forces within the Earth have caused many layers of sediment to tilt. Mountain ranges have formed as the tectonic plates that comprise Earth's surface have crashed slowly into one another. Erosion has worn ancient mountains flat, and new ones have formed to take their place. As a consequence, seams of coal may tilt up or down; they may bend or abruptly come to an end only to begin again nearby. All of this geologic upheaval makes the process of mining coal more challenging.

Meeting the Demand
for Coal

There is an astounding quantity of coal in the world, an esti-mated 990 billion short tons (900 billion metric tons) according the U.S. Energy Information Administration (EIA). Two-thirds of this is located in just four countries: the United States (27 percent), Russia (17 percent), China (13 percent), and India (10 percent). The economies of the United States, China, and India are some of the world's fastest growing and most coal-dependent. World demand for coal is growing rapidly, and there is no end in sight. This chapter discusses factors affecting demand, and discusses technologies used to meet the demand for coal with special reference to the United States, one of the world's largest and most technically advanced coal markets.

Meeting the demand for coal has always entailed accepting certain environmental, social, and health costs. The second goal of this chapter is to enumerate how a coal-based economy affects the

Mountaintop removal entails dismantling the mountain to obtain the coal inside. *(Wikipedia)*

environment and the health and safety of the men and women who procure the coal on which so many depend.

THE SCALE OF DEMAND

World demand for coal is enormous and growing. The EIA has predicted that world coal consumption will increase 74 percent, from 5,900 million short tons (5,400 million metric tons) in 2004 to 10,300 million short tons (9,300 million metric tons) in 2030. The EIA predicts that China and India will double their consumption of coal during this time, and the United States' consumption will increase by about 50 percent during the same period. The situation is

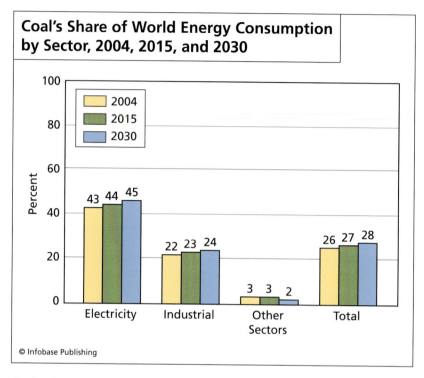

Coal's Share of World Energy Consumption by Sector, 2004, 2015, and 2030

© Infobase Publishing

Coal's share of world energy consumption by sector: Coal dependence increases over time. *(EIA)*

often more nuanced elsewhere. In Germany, for example, a number of new coal plants are planned or under construction, but many of these are envisioned as replacements for older, less efficient coal-burning plants. (Plans currently call for building more coal-fired plants to replace that nation's nuclear reactors, which are being shut down.) But Germany will not build coal plants to meet increased demand because Germany plans to limit future increases in the demand for electricity through conservation. That, at least, is the plan. In Canada the demand for coal will also grow at a slower pace than in the United States. The province of Ontario, for example, is working to find alternatives to its coal-burning plants as a pollution-control measure. Ontario plans to shut down its coal plants when those alternatives become available. But in western Canada additional coal-burning plants are planned in order to keep pace with the increasing demand for electricity. In what follows, most attention will be directed to the situation in the United States.

As previously mentioned, more than 90 percent of all coal consumed in the United States is used to produce electricity. As electricity usage increases, coal consumption tends to increase as well, but the relationship between electricity production and coal consumption is not a simple one. From 2004–05, for example, electricity consumption increased by 2.4 percent. Output from coal-fired power plants increased by 1.8 percent during the same time period. But coal consumption increased by 2.1 percent because power producers burned more low-sulfur coal, which happened to also have a lower heating value. (Because the heating value was lower, less heat was generated per ton of coal burned. Consequently, more tons of coal had to be burned to maintain electricity production.)

As of 2007, according to the EIA, there were roughly 80,000 workers engaged in the business of mining coal in the United States. This number fluctuates from year to year and even season to season. Over the short term, it depends on demand: As coal production increases so does the number of miners. Over the long term, however, the number of miners in the business has little to do with demand. Technology is much more important. Historically, productivity (the number of tons of coal produced per miner per hour) has increased faster than production, so, averaged over several years, as production has increased, the number of coal miners has decreased. In 1973, the average miner in the United States produced 2.16 short tons (1.96 metric tons) of coal per hour. Thirty years later, in 2003, the average miner was producing 6.95 short tons (6.30 metric tons) of coal per hour, an increase in productivity of 222 percent. The amount of coal produced per hour also depends very much on the method used to extract the coal, and that depends on the local geology.

Historically, the center of U.S. coal production was in Appalachia. For many years, coal mining was largely restricted to underground mines. During this time, coal was removed by a technique called *room-and-pillar* mining. To see how this works, imagine a coal seam as a large panel-like deposit of material lying more or less horizontally beneath the ground. Early miners could not remove all of the coal—or even most of it—because that would have created a large underground chamber with no supports for the roof. The roof would have collapsed on top of them. Consequently, they left huge pillars of coal unmined. These pillars, supplemented with supports installed by the miners, supported the weight of the earth above them. As a general rule, room-and-pillar mining entails leaving more than half of the coal in place. In 1950, when roughly 75 percent of all coal produced in the United States was extracted via room-and-pillar technology, the United States produced 560 million short tons (510 metric tons) of coal.

In the 1970s, another method of underground mining, called *longwall* mining, began to gain favor. While room-and-pillar mining can be used on deposits that are not horizontal or uniform in thickness, longwall mining makes essential use of these characteristics. Imagine a large rectangular panel of coal—typically on the order of 1,000 feet wide by 10,000 feet long (300 m × 3,000 m)—and imagine, further, that it is located deep underground, oriented horizontally, and of a fairly uniform thickness.

To use the longwall technique, it is necessary to first dig tunnels on all four sides of the panel. This is done with a machine that grinds coal out of the seam and drops it onto a conveyor belt that carries it out of the mine. This process, called continuous mining, leaves a tunnel in its wake. When the tunnels are completed, automated moveable roof supports are placed along the 1,000-foot (300-m) long face of the panel. Once the supports are in place, a machine, working beneath the moveable roof, begins to excavate coal, sweeping back and forth across the face of the panel. The coal falls onto mov-

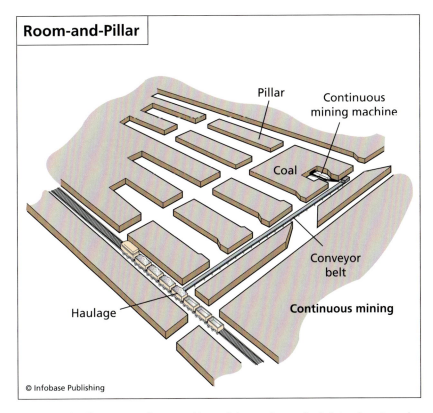

Room-and-pillar mining diagram. Most of the coal must be left in place in order to protect the miners. *(EIA)*

ing belts and is carried out of the mine. As the machine excavates more of the coal, it moves forward to maintain contact with the coal face. The automated roofing system also moves forward in order to protect the excavating machine from roof falls. As a result, a large pillarless room is created in the wake of the machine. The further into the panel the machine cuts, the larger the chamber it leaves behind. Eventually the ceiling of the room collapses, but the mining continues. Today, most underground mining is accomplished via longwall mining.

The alternative to underground mining is *surface mining.* This can be accomplished as long as the coal seam is not too deep un-

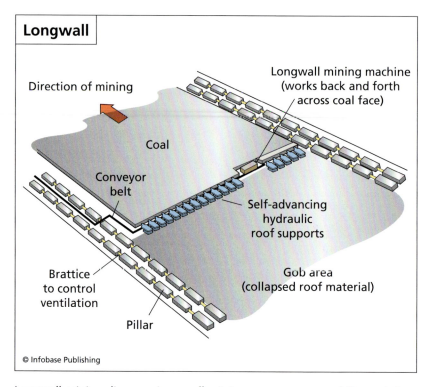

Longwall

Direction of mining

Longwall mining machine
(works back and forth
across coal face)

Coal

Conveyor
belt

Self-advancing
hydraulic
roof supports

Brattice
to control
ventilation

Gob area
(collapsed roof material)

Pillar

© Infobase Publishing

Longwall mining diagram. Longwall mining removes more of the coal than room-and-pillar mining, but as the roof collapses, surface water supplies can be disrupted and the foundations of buildings damaged. *(EIA)*

derground, and this is exactly the geology of the coal deposits in the western United States. To mine these deposits, the rock and dirt that lie above the deposits, called the "overburden," is stripped away and enormous machines excavate the coal from open-air pits. The key to increasing productivity in surface mines is to make larger pits and use larger machines. The scale of modern surface mining is hard to appreciate without seeing the mines in action. Gigantic excavators load immense, specially built trucks. All of this activity occurs inside deep pits that stretch far into the distance.

In the early 1970s, when the United States was producing almost 600 million short tons (544 million metric tons) of coal annually, production was evenly divided between surface and underground mines, but today, when production exceeds 1,100 million short tons (1,000 million metric tons), somewhat more than two-thirds of all coal produced in the United States is from surface mining. There are two main reasons for the shift. First, coal near the surface is simply easier and cheaper to obtain, and so surface mines tend to be much more productive—measured in tons of coal produced per miner per hour—than are underground mines. This is true even with the implementation of longwall mining technology and the increasing levels of automation practiced by underground-mine operators. Second, federal legislation, in particular the Clean Air Act of 1970 and its amendments of 1977 and 1990, made it more profitable to burn low-sulfur coal, and low-sulfur coal is generally located in the West where the largest surface mines are located. (Coal from Eastern and Midwestern mines generally has a higher average sulfur content.)

ENVIRONMENTAL AND SOCIAL COSTS OF MINING

One of the main environmental costs associated with mining is acid mine drainage. Some coal deposits contain significant amounts of iron sulfide, also called pyrite. The amount of pyrite depends upon how the coal deposit was formed, but as a general rule more is found in deposits in the eastern United States than in the West. When pyrite is exposed to the atmosphere it interacts with oxygen and water to produce acid. By itself, the acid is harmful to the surrounding environment, but it may also contain dissolved metals, such as nickel, aluminum, manganese, copper, or zinc, which compound the damage. How much acid is produced and the rate at which it is produced depend on the local chemistry, but acid production can continue for years or even decades after mining ceases. When this acid finds its way into surface waters, the result is water that is hazardous to humans and the broader environment. Minimizing acid damage has become an important goal of mine operators and regulators.

Currently, in the United States, about 10,000 miles (16,000 km) of rivers and streams are adversely affected by acid mine drainage. Most of this damage is associated with mines abandoned prior to the passage of federal legislation, especially the 1977 Surface Mining Control and Reclamation Act, created to regulate the effects of mining on surface waters. It is worth noting that 30 years ago about 18,000 miles (29,000 km) of streams and rivers were affected, so important progress has already been made.

Solutions to the problem of acid mine drainage are not cheap. Currently, the coal mining industry spends many millions of dollars each year treating acid mine drainage. Most of the problematic sites occur in the eastern United States, where local geology is more favorable to the production of acid. As understanding of the problem of acid production has improved, it has become increasingly possible to predict prior to the start of mining operations whether a proposed mine will produce substantial acid mine drainage. This enables regulators to either refrain from issuing a mining permit or to place additional restrictions on the permit that will ensure that acid mine drainage will be minimized. Prevention is, in this case, much cheaper than remediation.

A second important environmental concern associated with mining is the reclamation of the landscape after mining activities have ceased. Surface mining produces the most obvious changes in the landscape.

In the case of surface mines of the type found in Wyoming, beneath the overburden there are extraordinary seams of coal—an average seam in some areas is 65 feet (20 m) thick. To obtain the coal, the overburden is removed and a long pit or "box cut" is hewed into the earth. These can measure as much as 600 feet (180 m) wide and one mile (1.6 km) long. They are dug serially, one next to the other. As the overburden is removed from one pit, it is used to fill the previous pit. In this way, the landscape always consists of a relatively "small" mined area surrounded by reclaimed land and land that has yet to be mined. Care is taken to grade and reseed the land so that it looks as "natural" as possible.

There is another type of surface mining that is now practiced in Appalachia. Called "mountain top removal" mining, it is surface mining adapted to mountainous geology. As its name implies, mountain top removal entails finding coal deposits located in the tops of mountains. Often these deposits consist of several layers of coal folded into the mountain, one on top of the other and separated by non-coal material. The mine operator removes the overburden and intervening layers of non-coal material and deposits them as fill in a neighboring valley. The coal is extracted, and when the mining is complete, the flat surface formed by the remains of the mountain is "restored" in the sense that it can support vegetation and be used as a level surface for development.

Whether reclamation efforts can be classified as successful depends on one's expectations. For those who believe that the land should be restored to its original and pristine condition, such efforts can never be entirely successful because mining permanently changes the landscape. For others, a reclaimed landscape, properly done, is an opportunity. In Appalachia, for example, where many small towns are located in narrow valleys, a "reclaimed" mountaintop, which consists of a large level area, is sometimes described by local residents as a boon because it permits economic development that would not otherwise be possible. Other residents see the destruction of a mountain—and mountaintop removal does destroy the mountain—as the destruction of a landscape that they love. Naturally enough, they oppose the practice. There is, however, one other advantage to mountaintop removal that emphasizes the class and cultural differences between those engaged in the practice and those opposed to it: Mountaintop removal is less dangerous for the miners.

It is an inescapable fact that coal mining is hard on the land. The environmental effects just described (and others) can be reduced by paying close attention to the chemical and physical changes brought about by extracting coal, but many of the effects can never be eliminated. So far, society has tolerated the tradeoff between environmental disruption and energy production because burning coal is one of the cheapest and most reliable ways of generating large amounts of electricity. But there are other costs—human costs—that are associated with the coal business, and these must also be acknowledged in order to fully appreciate the cost of the United States' dependency on coal.

Understanding the health costs associated with mining coal is also a useful way of determining the cost of coal. More lives have been lost in the production of coal than any other primary energy source. The numbers are staggering. Two sources of mortality are considered here: black lung disease and traumatic injury.

The first problem in discussing any occupational disease is to demonstrate that it is, in fact, caused by the occupation in which the affected individuals are employed. A cause-and-effect relationship between work and illness is often difficult to establish because most diseases look the same no matter the cause. Some cases of lung cancer are associated with work—it is estimated that 16 to 17 percent of all lung cancer in men and 2 percent of all lung cancer in women are work-related—but whether a *particular* case of lung cancer was caused by the conditions experienced by an individual at work is often impossible to determine. Another difficulty in determining cause-and-effect relationships is that some diseases appear many years after exposure to the causative agent, and there may be several contributing factors of which only some are work-related. There is, however, one disease—coal workers' pneumoconiosis (CWP), also known as black lung disease—that is absolutely work-related, and, in fact, CWP is recognized by the federal Centers for Disease Control and Prevention (CDC) as an illness that is *caused* by working in coal mines.

CWP is caused by inhaling coal dust over prolonged periods of time. Coal dust lodges in the lungs and eventually damages the lungs and reduces lung function. A healthy lung looks pink; the lung of an individual with CWP has many black splotches on it. Between 1968 and 2008 approximately 75,000 deaths have been attributed to CWP, according to the information gleaned by the CDC. The prevalence of CWP has declined steadily since 1970 when an estimated one-third of all miners displayed the symptoms. The reason is federal legislation: In 1969 Congress passed the Coal Mine Health and Safety Act. This act established limits on dust exposure within underground mines with the ultimate goal of eliminating CWP. The limits established in 1969 have been reduced still further during the intervening years.

Because of intervention by the federal government, dust exposure and the incidence of the disease were reduced, but CWP has not been eliminated. Today, CWP occurs in miners who have spent their entire professional lives working under the standards established by the federal government, leading many to conclude that the standards are not yet strict enough. According to the CDC, the last year that CWP deaths exceeded 1,000 was 1999. Mortality from CWP is declining very slowly.

Accidental injuries are also a significant source of mortality. During the 20th century, the most coal miners to have died in a single year in the United States as a result of traumatic injury occurred in 1907 when 3,242 miners died. (As previously mentioned, 100,000 miners died in mining accidents during the 20th century and many more died of CWP.) Until 1930, it was more the rule than the exception that in excess of 2,000 miners died per year in mining accidents. Since 1947, the number of fatal accidents has trended downwards.

Not since 1984, when 125 miners died, have more than 100 U.S. coal miners died in accidents in a single year. Today far fewer

(continued on page 46)

An Interview with Charlene Marshall on the Human Costs of Coal

Charlene Marshall is a lifelong resident of Monongalia, West Virginia, and is a graduate of the segregation-era public school system. She attended Bluefield State College and was married to the late Rogers Marshall. They had three children. For 15 years she worked at Sterling Faucet in Morgantown, where she was active in the union, serving as recording secretary for Steelworkers Local 6214. She later worked for the West Virginia Department of Weights and Measures and as staff at West Virginia University. In April of 1991, she was elected to the Morgantown City Council, and in July of that same year, she was elected mayor of Morgantown, the first African-American woman elected as mayor in West Virginia. She served as mayor for seven years. She is now serving her fourth term in the West Virginia House of Delegates. She has received numerous awards and is past president and lifetime member of the Morgantown chapter of the NAACP. The following interview took place April 3, 2008.

Charlene Marshall *(Courtesy Charlene Marshall)*

Q: Tell me a little about your early background in West Virginia.

A: I was born about four or five miles outside of Morgantown in a little coal-mining town called Osage. I had six siblings.

My father was a coal miner. He was killed in the mines when I was real young, leaving our family with five children, and my mother was expecting a child. (My mother also lost her father when she was a child, as a result of a mine accident.) It was pretty rough growing up.

A few years later my mother married again, and my stepfather was a coal miner, and then we had one additional sibling, a sister. And when my little sister was four years old, my stepfather was killed in a mine.

At that time, I had just gone away to school, and, of course, I had to return home. Then I married. My husband lived in Morgantown. That is where he grew up, and so I moved a few miles to Morgantown. That's where my husband and I raised our family.

Q: And about your father and stepfather—

A: My father was killed in a slate fall in a single accident. My father was killed in October. I was five or six years old. He was killed in a mine very close to where we lived in Osage. As I said, that was in October, and in May there was an explosion in that mine, and a number of miners were killed in an explosion in that same mine. My stepfather oddly enough was also killed in October, and it was a single accident. He drove the motor, taking miners into the mines. His motor jumped the track—I guess as he went in—near where they set the beams in the mines. It went some distance and tore some of those beams out, and I remember that his face was distorted so much that they had to use a photo to rebuild one side of his face.

Q: There are two types of costs associated with mining, safety and environmental. During the 20th century about 100,000 American miners were killed in mining accidents. That doesn't count the higher number of miners who died of black lung disease.

A: Yes.

Q: Mining is still a dangerous profession, but it has gotten safer. You are a member of the House of Delegates Select Committee on Mine Safety—

A: Yes. I'm vice-chair, and safety has gotten better.

(continues)

(continued)

The Committee went into the mines in September, I think it was, of '07. At first, I said I wouldn't go, and the Chairman said, "Oh, I'll take care of you. You'll stay close with me," and so forth. So I decided to do that, and maybe get over some fears that I had. And I did go in the mine. That was my first time. We were probably in the mine about four hours. We went down 1,000 feet, and we probably traveled two different sections and went back into the mines approximately eight miles.

Members of my family—I didn't let them know that I was going to do this, and they didn't know it. For instance, my daughter didn't know that I had been in the mines until she read it in the paper. And everyone in the family—their reaction was pretty much the same: "You did what?" But I just wanted to do that the one time. It was Mine Safety, the committee, that went in. It squashed some concerns and fears that I had, and I was really amazed. It was a lot different than I thought it would be.

Q: How was it different?

A: Actually, the areas that we traveled in—there was more space than I thought there would be anyplace in the mine.

Q: Overhead space?

A: Overhead space and even the width of the mine. There was more overhead space, and I expected everything to be black and dreary. I had heard about rock dusting before. We had a couple of friends a few years ago that did rock dusting—

Q: I'm sorry Ma'am. What is rock dusting?

A: Rock dusting—I don't know exactly what type of a powder you would call it, but it's real bright. It's just white. It reminds you of limestone. You know what that is?

Q: Sure.

A: And of course that's white. They do this dusting to avoid explosions.

Where I grew up, you would just see an opening in the side of the hill, and that's the way that the miners would enter the mines. Later, when there were new mines built, I could never figure out, where was the opening? Well, the newer ones, they sink a shaft, and that's how they get in. So this one where we went in, you got on an elevator and went down, so from the outside, you would never have any idea how they entered the mine. We went down on the elevator, and once you exit the elevator there was a short hallway built with cinderblocks on both sides. You went through there a few feet, and you were inside the mine.

Then we got on the motor to take us further in. We did get to where a longwall was being operated. Where they were working, that part had not yet been rock dusted, so that was black like I had imagined it would be. I imagine—and I don't know this to be exact—but I imagine that a rock dust only occurs once they have worked the area and moved on.

Q: In addition to the costs that many of the people in coal mining regions have suffered in terms of injuries and deaths, there is also the question of environmental costs—damage to water supplies, mountaintop removal, and so forth. Do you think that the environmental situation associated with mining has also gotten better?

A: I would say possibly, in some areas. I don't truly know the answer to that. The reason I'm hesitating is that I've often wondered about whether when a mine is closed and they move on, whether they do enough to ensure that the recovery is great.

Where I grew up . . .

Q: In Osage?

A: Yes, in Osage. There was a small stream that went through that little town, and I can remember wondering, when I was a little child, why that little stream would from one day to another be a different color. It would be a brownish color. It would be orange, and individuals would say, "It's because of mine drainage." Well, I think that that stream still has different colors to it. I don't know what's going on there and what is causing it, but I believe that people are now more careful to avoid children playing in these streams and so forth. We were never allowed to do that, but I don't

(continues)

(continued)

know if people realized—I don't know how many little children waded in that stream before, and it goes for miles.

Q: I've read that spread out across the coal regions of Appalachia there are about 5,000 miles of streams that have been affected by acid mine drainage. Some say the number is much higher.

A: I didn't know the distance or whatever, but I have just often wondered what are the lasting effects of those streams? That particular one—that was the only one I was aware of as I was growing up.

Q: By lasting effects, do you mean the lasting effects on the landscape or the lasting effects on the people who live nearby?

A: The people who live nearby.

Q: Part of the problem, I think, is that there is no easy way to remove hundreds of millions of tons of coal from the ground each year, and hundreds of millions of tons must be removed. About 50 percent of the electricity in the country is generated from coal.

A: Correct.

Q: So most people in the country share in the benefits of coal whenever they use electricity, but most people don't share in the costs.

A: Correct. When you mention the costs, I think about the costs. Number one, the cost to individuals. How does it affect their health in later years? And the costs as far as some of the accidents that have occurred—the costs because of those deaths and the ongoing lasting effects on those families.

When I was born, it was just men, so you tend to think of the costs of those fathers not returning home, but now it's mothers and fathers, men and women. The costs of— As a child, in the area where I lived it was not unusual to see—not a large number—but a number of men with one leg. Of course, back in those days they had a peg leg that they put on. It was not a lot (of men), but one is too many. I can remember some of the individuals that had a leg crushed in the mines, and it was amputated. The cost is so high in a number of ways.

Q: Yet, people are not aware of it outside of—.

A: —outside of this area. And I don't know how far the area extends where people aren't aware of it. I imagine that maybe it extends into the cities and so forth.

When we think of what the coal provides . . . to me it's almost like . . . for those individuals (who aren't aware), there should be a way to make them know about what they are receiving and at the cost that they're receiving it. It sort of irritates me because they think of these individual (miners) as being just—I guess so many times they're belittled in such a way. I know that some of these individuals—if they were fortunate enough to escape accidents, fortunate enough to work a number of years in the mines and come out and be able to retire—some of them have had good lives. But still there are so many individuals away from here who belittle them and look down on individuals from West Virginia.

Q: It's true. What future work do you think needs to be done to improve the situation in West Virginia with respect to coal and coal miners? What is left to be done?

A: Although we've made some improvements, no matter what you do for improvements, there is always room for additional work, additional safety issues. It's just that so many times it's only after an event—maybe something terrible happens—that we realize that we should have taken care of this. So I just think that we need to continue to observe where we can go for improvement, and I always think of the benefits for the families.

As I was growing up, the benefits were terrible—

Q: Did your family receive anything when your father and your stepfather died?

A: I don't remember the amount from the time that my stepfather died, but possibly—and I've thought about this so many times although I didn't think about it as a youngster—but by the time my stepfather was killed, my younger siblings were still young enough to be receiving benefits

(continues)

(continued)

from my father. Those benefits at that time were very low. For the widow, either 30 or 50 dollars a month and five dollars for each child.

Well, I have no idea what the benefits are today and I know there's social security, but I've just been someone who has been in a union. If I was in a union today, I would want to fight for better benefits. I would not want anyone to go through some of the things that those individuals who came along when I did—that we went through.

Q: Thank you very much for sharing your story.

(continued from page 39)

miners die in mining accidents each year than was the case even a few decades ago—usually between 30 and 50 accidental deaths occur—although it is also true that the number of miners is much smaller. The workforce in 2006, for example, is 57 percent as large as it was in 1984, and the number of fatalities in 2006 (47) is 38 percent as large as it was in 1984, so the number of fatal accidents is trending downward somewhat faster than the number of miners. (Outside of developed countries, coal miners frequently work in situations that would have been familiar to their 19th-century predecessors in the United States. In China, for example, thousands of coal miners die each year.)

It has always been true—and it remains true—that the tragedies that routinely occur to coal miners and their families rarely make news reports, and one can only wonder why. It is worth emphasizing that most of these deaths are preventable. CWP, for example, is known to be caused by inadequate ventilation, and methods of maintaining good ventilation are well understood. With respect to coal mining accidents, the one accident to receive widespread media attention in 2006, the Sago mining accident in which 12 miners died, was, according to the United Mine Workers of America (UMWA), both foreseeable and preventable. The 2007 Crandall Canyon mine accident, in which nine miners died, was also preventable, according to the UMWA. The number of deaths caused by mining coal is still one more way to estimate the cost of humanity's reliance on this fossil fuel. The human price paid for access to inexpensive coal remains very high.

4

The Combustion Reaction

Combustion is a chemical reaction that combines materials called *reactants* to produce materials called products—plus heat. The products produced by the combustion of coal include water, carbon dioxide (CO_2), and numerous pollutants such as sulfur dioxide (SO_2). But the purpose of burning coal is to produce heat; the products are, in this sense, secondary. To understand the consequences of burning coal on an industrial scale, one must acquire some familiarity with the mechanics and chemistry of the reaction as well as some of the principal effects that coal combustion has on the broader environment. Those are the goals of this chapter.

PRODUCTS AND REACTANTS

Burning large amounts of coal presents difficult technical challenges, not least of which is that much of what is found in raw coal should not be burned or cannot be burned. As described in chapter

Although sometimes described as an antiquated fuel, coal is a vital primary energy source for some of the world's most advanced economies. *(Courtesy NetRegs)*

2, unprocessed coal, coal straight from the mine, contains ash, sulfur, and other materials. Ash does not burn, but it can present an air pollution problem. Sulfur, which does burn, poses still another set of problems (see chapter 2). Therefore, if a coal supply has a high ash or high sulfur content it is often processed before it is burned in order to remove as much of the ash and sulfur as possible. This type of processing also assures the power producer that the supply of coal it is burning is reasonably uniform. A uniform fuel supply enables the power producer to predict how much thermal energy and how much pollution will be generated per unit mass of fuel burned.

Removing ash and sulfur from coal prior to combustion is often accomplished by a procedure called washing. Coal washing techniques are designed to make use of the fact that carbon is less dense and consequently more buoyant than ash and sulfur pyrites. As a consequence, when suspended in the right liquid, the ash and pyrites will sink—that is, they separate from the carbon part of the coal and leave behind a "purer," less polluting fuel. The washing process begins by crushing the coal. Next, the coal pieces are separated by size to match the capacities of the various coal washing devices. Third, the coal is passed through coal washing devices, where it may encounter liquid-filled spinning containers, pulsating streams of water, or even liquids with densities different from water, all with the purpose of floating the coal and sinking the impurities. Finally, the cleaned coal is separated from the liquid. (What remains behind is a liquid full of undesirable materials. This has sometimes found its way into the water table and ruined the drinking-water supplies of those living near coal mines. In any case, this slurry always presents a significant waste disposal problem.)

The coal is not yet ready to be burned. Coal must be further processed, but the way that it is processed depends on the design of the furnace in which it will be burned. One standard method of burning coal requires that the coal be pulverized prior to being introduced into the combustion chamber. Pulverizing coal, which only changes the form (not the chemistry) of the coal, is important because burning is a surface phenomenon—when a chunk of coal is burning only the material at the surface is on fire. The carbon inside the chunk cannot burn because it has not yet been exposed to oxygen. In order to release all of the chemical energy in a piece of coal quickly, it is necessary to reduce the coal to a powder so that all of the coal is very near the surface. When a sample of pulverized coal is exposed to a fire, the coal particles will burn virtually instantaneously. This is what happens in a pulverized-coal–fired burner. Power plants using this technology are common.

Another technology for burning coal is called *fluidized bed combustion*. This technology is not nearly as widespread as pulverized coal combustion, but it has found a market among power producers who anticipate cofiring coal with *biomass*. (Cofiring means burning coal together with another non-coal fuel.) Coal prepared for a fluidized bed does not need to be crushed as finely as that for a pulverized-coal–fired furnace, but the particles are still fairly small—less than 0.25 inch (0.6 cm) across. These particles are introduced into a combustion chamber. Simultaneously, a powerful and continuous stream of air is blown upward from the bottom of the combustion chamber to keep the fuel suspended in the air. The situation is sometimes compared to a rapidly boiling pot of soup. Other materials can be introduced into the burning turbulent mass as well—limestone, for example, is introduced because it helps to remove sulfur dioxide from the *products of combustion*—and the materials burn, mix, and react while they are consumed in the fire.

So far only the mechanical techniques for processing and burning coal have been described. Understanding the importance of the combustion reaction is also important, but combustion is a complex chemical phenomenon and coal is a complex fuel. Only the most important reactions are described here.

Carbon (chemical symbol C) and oxygen (chemical symbol O) are the main reactants when coal is burned. The oxygen comes from the air that is blown into the combustion chamber. But oxygen in the air exists as a molecule consisting of two oxygen atoms chemically bound together. Its chemical symbol is O_2—the O represents the oxygen atom and the subscript 2 indicates that two oxygen atoms are bound together to form the oxygen molecule. The combustion reaction occurs when an atom of carbon combines with a molecule of oxygen. The result is a molecule of carbon dioxide (CO_2), which consists of the single carbon atom bound to the two oxygen atoms, plus heat. This long description can be summarized in the much simpler-looking chemical equation:

$$C + O_2 \rightarrow CO_2 + heat$$

Other chemical reactions also occur. *Hydrocarbons,* a class of molecules consisting of various numbers of carbon and hydrogen atoms, are also present in coal. Combusting hydrocarbons produces CO_2, H_2O (water)—plus heat. Other materials present in coal also react with oxygen. Sulfur, for example, combines with oxygen to produce SO_2. Nitrogen may combine with oxygen to produce nitrogen oxides, a group of reactive gases, one of which, nitrogen dioxide, can sometimes be seen as a reddish-brown layer of pollution suspended in the air above cities. Nitrogen oxides can cause respiratory problems and contribute to acid rain. Many other chemical reactions also occur during the combustion process, reflecting the fact that coal is a chemically complex substance.

There are also other materials that do not burn but are released from the coal during combustion. Sometimes they are captured by antipollution equipment and sometimes not. Mercury is an example of this type of material. It is often present in trace amounts in coal. When the coal is burned, some of the mercury is released into the combustion gases and carried up the smokestack and out into the atmosphere. (How much mercury is released in this way depends on the technology used to process and burn the coal.) The mercury emitted from the smokestack eventually settles on Earth's surface. Many freshwater fish living in mountain lakes in the northeastern part of the United States now carry trace amounts of mercury that originated in the coal-fired power plants of the Midwest. Because mercury accumulates in the body of the organism that ingests it, these fish would be hazardous to any organism, human or animal, that made a regular diet of them. Eating such a fish means ingesting a small amount of mercury. As the mercury accumulates, so do the health risks; young children, in particular, are at increased risk of neurological problems when exposed to mercury.

Most of what remains after coal is burned is, however, more remarkable for its abundance than its toxicity. Called coal combustion products (CCP), U.S. coal-burning power plants produce about 120 million short tons (110 million metric tons) of CCP each year. About 57 percent of all CCP is fly ash, most of which is removed by electrostatic precipitators, antipollution equipment designed for that purpose. Fly ash is used in cement. Heavier material called bottom ash constitutes about 15 percent of CCP, and much of that finds its way into concrete. Most of the rest is material captured from the combustion gases by a piece of antipollution equipment called a flue-gas desulfurization unit, or is itself a product of the desulfurization process, and some of this material is used in making wallboard. Not surprisingly, more than half of all CCP finds no use, because there is too much CCP to use—another testament to the scale at which coal is burned in the United States.

GLOBAL WARMING AND CARBON SEQUESTRATION

Global warming is the phenomenon by which changes in the chemistry of Earth's atmosphere cause increases in Earth's *average* temperature. Earth's average temperature is a statistical quantity obtained from the analysis of numerous widely separated measurements. That Earth's average temperature increases does not mean that temperatures at a particular location will increase. During the process of global warming, some regions may well cool as others warm, because rising temperatures will also affect the way that thermal energy is distributed across Earth's surface by atmospheric and oceanic currents. And even if the average temperature increases, there will still be colder years and warmer years just as there have always been, and so one or even several unusually warm years do not indicate that the planet is becoming warmer on average anymore than one or several unusually cool years indicate an approaching ice age. Care must be taken in interpreting the evidence of one's senses. The available statistical evidence, however, indicates that Earth's average temperature is increasing, that it has been increasing for some time, and that humankind is the cause.

But if the effects of an average increase in temperature are complex, the mechanism by which temperatures increase is less so. As the Sun shines on Earth some of its energy is reflected back into space by clouds and dust, but most of it passes through the atmosphere in the form of light. Earth's atmosphere is largely transparent to light. Consequently, light passing through the atmosphere causes only a small increase in atmospheric temperature. At Earth's surface, some light is reflected back into space, and some light is absorbed and then radiated back into the atmosphere as heat. Heat, also called infrared radiation, does not pass as easily through the atmosphere as light. Some of the heat is retained by the atmosphere in a process called the greenhouse effect. (On a sunny day, energy will pass through the glass of the greenhouse in the form of visible light; some of the light energy is absorbed by plants and equipment inside the greenhouse and radiated back into the air of the greenhouse in the form of heat. Because heat cannot pass as easily through the glass walls of the greenhouse as did the light, it begins to accumulate, causing an increase in the temperature inside the greenhouse.)

The atmosphere assumes the same function as the glass in the greenhouse—hence the name of the phenomenon—and just as one can change the heat-retention properties of glass by changing the materials of which the glass is made, one can change the heat-retention properties of the atmosphere by changing its chemical composition. One gas that has good heat-retention properties is carbon dioxide (CO_2), a gas that is naturally present in the atmosphere. An increase in atmospheric CO_2 levels will, all other things being equal, cause Earth's atmosphere to become more efficient at retaining heat, and average global temperature will rise.

A commercial coal-fired power plant will burn many tons of coal, which is primarily carbon, each day, and consequently it will produce many tons of CO_2 each day that it is in operation. All of the CO_2 so produced finds its way up the smokestack and into the atmosphere. Some CO_2 will be quickly absorbed by plants or by the ocean, but some CO_2 will circulate in the atmosphere for longer periods of time. As a consequence, atmospheric CO_2 levels are now higher than they would have been had humanity not turned to fossil fuels, especially coal, as primary energy sources. The concentration of CO_2 in the atmosphere continues to increase.

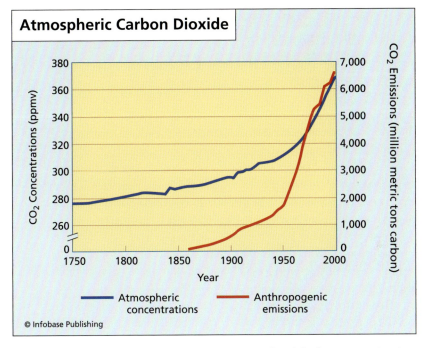

As the rate of carbon dioxide emissions due to fossil fuel consumption has soared, so has the concentration of carbon dioxide in the atmosphere. *(Oak Ridge National Laboratory)*

So far no nation with large coal resources has been willing to stop burning coal to produce electricity. Coal is inexpensive and too plentiful to ignore as a potential energy source. But even though burning coal must result in the production of CO_2, it does not follow that the CO_2 must be released into the atmosphere. Engineers have begun to experiment with the *sequestration* of CO_2. Carbon dioxide sequestration generally involves capturing CO_2 at the point at which it is produced and then storing it more or less permanently. (See "More on CO_2 Sequestration" in chapter 5.) In this chapter, only the broad concept of sequestration is described.

While there have been very limited attempts to use sequestration technology, virtually all of the electricity currently produced by coal-fired power plants is produced without the use of sequestra-

The Role of the Oceans

Today, some nations—most notably the United States and China—burn enormous quantities of coal for electricity or to provide heat for industrial processes. So much coal has been burned and so much CO_2 has been vented into the atmosphere that the amount of CO_2 in Earth's atmosphere has measurably increased. Billions of tons of CO_2 are vented into Earth's atmosphere each year, much of this due to coal-fired power plants. But the *measured* increase in atmospheric CO_2 levels is less than the amount of CO_2 actually emitted. What happened to the rest of it?

Scientists often speak of "sinks" and "sources" of CO_2. A source is a body or a process that produces CO_2 and a sink is a body or process that absorbs the gas. Oceans function as sinks. They currently absorb about one-third of all CO_2 emissions. The primary mechanism by which this occurs is a phenomenon that takes place near the oceans' surface, in which CO_2 is absorbed directly from the atmosphere. In the North Atlantic, the water that absorbs the CO_2 is convected northward until it reaches an area where currents descend from the surface into the ocean's depths. The largest of these currents is called the thermohaline current, and it flows from the Arctic southward along the deep ocean basin. Eventually, it rises again to the surface. The thermohaline current travels slowly over large distances. Consequently, the absorbed CO_2 remains in the ocean for several centuries, but some of it is eventually released back into the atmosphere when the thermohaline current resurfaces. The process by which CO_2 is absorbed in this way is inorganic—that is, no living creatures are involved.

Inorganic processes are supplemented by biological processes. Many small sea creatures require carbon to grow. They obtain carbon from the water around them, and as they absorb this carbon, they decrease the concentration of carbon in the ocean's surface waters. From the point of view of those interested in climate change this has two effects: First,

tion—that is, almost all coal-fired power plants vent all their CO_2 directly into the atmosphere.

What makes sequestration attractive is that with sequestration it seems as if one can continue to burn coal without impacting the provided the creatures sink to the bottom when they die, some carbon is again removed from the atmosphere, and second, by decreasing the concentration of carbon near the surface, the ocean is again able to absorb more atmospheric carbon through the inorganic process already described.

Some researchers have proposed fertilizing the ocean in order to stimulate the growth of ecosystems that inhabit the near-surface regions of the ocean. If near-surface biological activity increased, the rate at which the oceans absorbed carbon would also increase. But making a measurable impact on the rate at which the oceans currently absorb CO_2 would require fertilizing vast expanses of water. Furthermore, if only some nutrients are added—iron is often suggested as a nutrient that if added in sufficient quantities would cause a quick bloom of single-celled organisms—other nutrients might quickly become exhausted. Growth is always limited by the availability of the scarcest essential nutrient, so successfully fertilizing the oceans may prove more difficult than the proponents of this method of climate stabilization seem ready to admit. Further research is needed.

Another proposed method of enhancing the role of the oceans as carbon sinks is to inject carbon directly into the ocean. There are several methods of injection under consideration, but the process starts at power plants, where most coal is burned. The CO_2 is captured before it is emitted into the atmosphere and transported to the ocean. It is then injected deep beneath the surface where, at the high pressures that exist at depth, it would change phase, becoming a liquid, and forming pools of CO_2 at the bottom of the sea. These processes enable the user to "hide" the CO_2—if not permanently, then at least for centuries.

global climate. Sequestration technology is not perfect, but capturing 90 percent of the CO_2 produced is certainly a reasonable goal for a commercial unit, and a 90 percent recovery rate would have a significant impact on global CO_2 emissions.

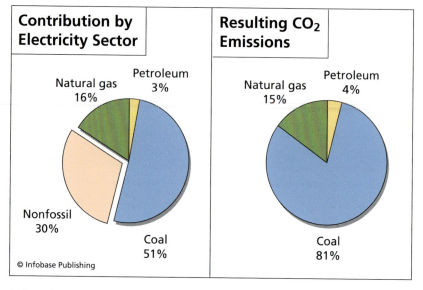

Contribution by Electricity Sector	Resulting CO$_2$ Emissions
Petroleum 3%	Petroleum 4%
Natural gas 16%	Natural gas 15%
Nonfossil 30%	
Coal 51%	Coal 81%

© Infobase Publishing

Although a very important energy source in the United States, coal is an even greater source of carbon dioxide emissions. *(EIA)*

There are several difficulties with sequestration, not the least of which is that the technology is expensive, limiting its commercial potential. Roughly speaking, however, the main barriers can be summarized with two observations: First, the volume of CO$_2$ produced by coal-fired power plants is enormous; consequently, any storage technology would have to be operated on an equally enormous scale. None of this infrastructure currently exists. Second, although it is possible to remove CO$_2$ from the products of combustion, the process requires expensive equipment, and it is energy-intensive. In order, for example, to produce 500 megawatts (MW) for sale on the grid, a coal-fired plant using sequestration technology would have to be made larger and burn substantially more coal than a plant without sequestration technology—perhaps as much as 25 percent more if outfitted with current sequestration technology—in order to produce enough energy to capture its own CO$_2$ and still have 500 MW left to sell on the market.

Some countries—including the United States, China, India, and Germany—plan to generate a large fraction of their future electricity requirements with coal with or without sequestration technology. The latter three countries are, in fact, already engaged in a coal plant-building boom, and none of these new plants, which will operate for decades, use sequestration technology. Sequestration technology will be expensive to install and expensive to operate should it prove to be practical at all. What sequestration technology offers—at a significant cost—is the possibility of meeting energy needs while reducing the impact of combustion on the global climate. Whether this proves sufficient incentive for the owners of the next generation of coal-burning power plants is not yet clear.

Electricity Production and Its Consequences

*P*ulverized coal technology is more than 60 years old. (This technology was described in chapter 4.) Reliable, relatively inexpensive to build and operate, and capable of generating power on a large scale, pulverized coal technology has been the standard technology for large-scale power plants for decades. This has begun to change in recent years as engineers have sought to respond to two conflicting imperatives: the demand for more electricity and the demand for reductions in power-plant emissions. Today a number of new and creative coal-burning power-plant designs are in limited use. The first goal of this chapter is to describe two of the best new designs. Creating technology to reduce emissions other than CO_2 from old and new coal-fired power plants is a lively area of research. Describing some of these technologies is the other goal of this chapter.

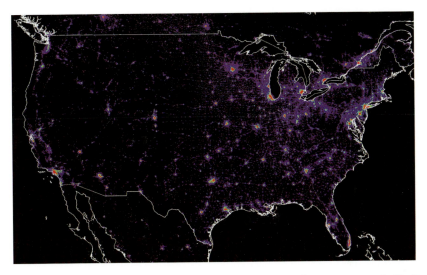

The contiguous 48 states at night—coal-fired power plants generate half of the electricity in the United States. *(NASA)*

NEW POWER-PRODUCTION TECHNOLOGIES

All coal-fired power plants are *heat engines* that convert thermal energy into electrical energy. In a sense, there is nothing special about coal-fired power plants; the same general principles that describe their operation apply to other heat engines, such as natural-gas–fired plants and nuclear plants. First, thermal energy is produced—in this case, by burning coal—and this energy is transferred to what is called the working fluid, usually water. The water turns to steam. The expanding steam drives a machine called a steam *turbine*. Steam turbines are mechanical devices that convert the linear motion of the expanding steam into rotary motion. Attached to the turbine is a device called a generator, which converts the rotary motion of the turbine into electricity. It is a simple-sounding process: Thermal energy flows from the heat source to the water causing it to change from liquid to steam; the steam drives a turbine; the turbine drives the generator, and the generator produces electricity.

After passing through the turbine, the steam is converted back to liquid water in a device called a condenser, and the liquid water is pumped back to the heat source so that the entire cycle can begin again. What distinguishes coal-fired plants from other types of heat engines is that the fuel is coal, the burning of which produces more harmful emissions than any other type of fuel.

One method of reducing emissions is to improve plant *efficiency*—that is, to produce the same amount of electrical power while burning less fuel. Burning smaller amounts of fuel produces fewer pollutants. Consequently, even if no other changes are made, a more efficient plant will emit less pollution per unit of coal consumed. Standard pulverized-coal–fired plants convert about one-third of all the thermal energy they produce into electrical energy. But this means that the remaining two-thirds is wasted—wasted in the sense that it is not converted into electricity. The plant operators paid for all of the chemical energy in the coal, but they failed to convert two-thirds of it into electricity. For decades this was not considered a problem. Faced with an endless supply of cheap coal and few concerns about power-plant emissions, power producers had little incentive to change. Today, power producers are faced with increasingly stringent antipollution regulations, and plant efficiencies are on the rise.

It may seem that the goal of every power-plant designer should be to design a plant that is 100 percent efficient in the sense that the plant converts 100 percent of all the thermal energy it produces into electrical energy. But it is a fundamental principle of science that it is impossible for any heat engine to convert 100 percent of its thermal energy into electricity. Instead, every plant has an upper limit on how efficient it can possibly be, a limit determined solely by the difference between the operating temperature of the plant and the temperature of the environment. It is, of course, always possible to build and operate a plant that is less than optimally efficient, but it is impossible to build a plant that exceeds this limit.

One higher efficiency design is the so-called supercritical pulver-ized-coal–fired plant. Supercritical plants convert a higher percent-age of the thermal energy that they produce into electrical energy. (The term *supercritical* refers to the working fluid. Recall that under normal conditions water is a solid [ice], or a liquid, or a gas [steam]. But when placed under high enough temperatures and pressures, water becomes "supercritical" and simultaneously displays some of the properties of a liquid and some of the properties of a gas. It is not that some of the water is liquid and some is steam, as is the case in a kettle of boiling water. Instead, all of the water assumes a sort of "in-between" phase that exhibits some of the properties of a liquid and some of the properties of a vapor.) A supercritical power plant is built to operate at temperatures and pressures high enough that the water that is used as the working fluid becomes supercritical. By operating at these higher temperatures and pressures, the plant produces the same amount of power as a traditional pulverized-coal–fired plant, but with significantly less fuel.

Early attempts to build supercritical plants were frustrated because the metal alloys that were used to build the plant com-ponents quickly corroded when exposed to supercritical water. These problems have now been solved, and modern supercritical plants located in the United States, Europe, and Japan operate at efficiencies exceeding 40 percent. From the plant operator's point of view, higher efficiencies are valuable because less fuel needs to be purchased to generate the same amount of power as a lower efficiency plant. Consequently, the plant owner is compensated for some of the higher costs incurred in building a supercritical plant by the savings in fuel. In addition, because supercritical plants burn less fuel, they also create fewer pollutants per unit of energy produced. Therefore, all things being equal, a more efficient plant also pollutes less. Supercritical plants are "evolutionary"—that is, they are a logical extension of the old pulverized-coal–fired plant designs.

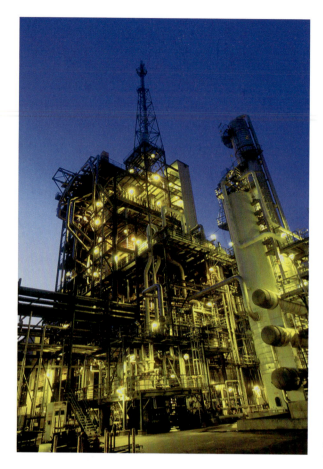

Wabash River IGCC Plant. This power plant uses the most advanced coal-based technology available. *(SgSolutions)*

A more radical departure from conventional coal plant design is the integrated gasification combined cycle plant, better known as the IGCC plant. Strictly speaking, they do not burn coal at all; they burn a gas obtained from coal. The process begins in a device called a *gasifier,* which heats coal to high temperatures with strictly controlled amounts of oxygen and steam. The result is a combustible gas, together with a number of pollutants such as sulfur dioxide, mercury, carbon dioxide, and other environmentally harmful materials. Next the gas is cleaned. The pollutants are removed, and what remains is the fuel. (Because it is easier to clean the reactants than it is to clean the products of combustion, more of the pollutants can be removed.) In theory, therefore, IGCC plants produce fewer pollutants.

After the gas has been cleaned, it is used to drive a gas turbine. The working fluid for a gas turbine is the combustible fuel that was produced in the gasifier plus the air with which it is mixed. The fuel-air mixture is compressed and ignited. The heated gases expand and push against the blades of a turbine causing it to spin; the turbine drives a generator, and the generator produces electricity.

The conversion of some of the thermal energy of the combustion gases into the rotational energy of the turbine causes a drop in the temperature of the combustion gases, but even with the reduction in temperature the exhaust gases remain very hot. Consequently, they are used again—this time to convert liquid water into steam. The steam is used to drive a secondary (steam) turbine just as is done in a conventional coal-fired plant. This is the "combined cycle" in the IGCC system: The gas turbine cycle is combined with the steam turbine cycle. Combining two cycles in this way makes IGCC plants more efficient in the sense that they make more complete use of the thermal energy that they produce. In theory IGCC units can operate at better than 50 percent efficiency.

Despite their higher efficiencies and lower emissions, comparatively few IGCC plants are in operation. Most plants still burn pulverized coal. There are three main reasons. First, even a conventional pulverized-coal–burning plant is fairly expensive to build, and because power producers have already spent large sums of money constructing these plants, most are reluctant to close them prematurely. Second, IGCC plants are much more complex than either pulverized coal plants or supercritical coal-fired plants, and their greater complexity has made them more expensive to build and operate. Finally, IGCC technology has not yet been developed to the point where it is as reliable as its more conventional counterparts. Consequently, IGCC plants represent more risk to investors. Typical is Bill Fehrman, president of PacifiCorp Energy, who in

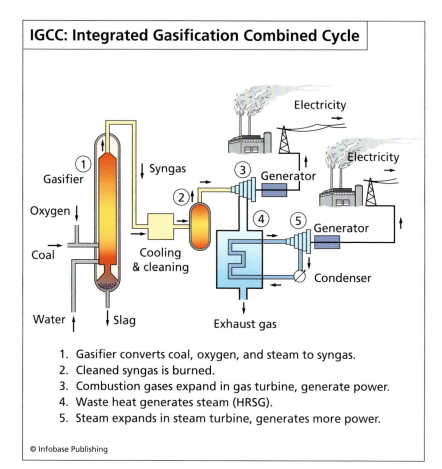

IGCC: Integrated Gasification Combined Cycle

Electricity

Electricity

① Gasifier

Syngas

③ Generator

②

Oxygen

Coal

Cooling & cleaning

④ ⑤ Generator

Generator

Condenser

Water ↑ ↓ Slag

Exhaust gas

1. Gasifier converts coal, oxygen, and steam to syngas.
2. Cleaned syngas is burned.
3. Combustion gases expand in gas turbine, generate power.
4. Waste heat generates steam (HRSG).
5. Steam expands in steam turbine, generates more power.

© Infobase Publishing

IGCC technology is currently the best hope of reducing the environmental impact of burning coal.

2007 testimony before the Senate Committee on Energy and Natural Resources, said, "With the [IGCC] technology unproven, with unclear costs, and with no guarantees from vendors, we are unwilling at this time to expose our customers to these risks."

Presumably there are workable engineering solutions to the problems of reliability, cost, and complexity, and many nations are currently spending billions of dollars developing better IGCC technology. Despite their current problems, therefore, IGCC plants will probably be the next step in coal-based power plant design.

More on CO₂ Sequestration

One technology for safely and "permanently" storing CO_2 involves underground storage. Injecting CO_2 into the ground is not new technology. What is new is the amount of CO_2 to be stored and the requirement that the storage be permanent.

For decades it has been common practice to inject CO_2 into old oil and gas wells. Old oil and gas fields still contain significant amounts of the resource, but recovery becomes increasingly difficult and expensive as the reservoir pressure (the force that originally caused the oil and gas to flow spontaneously up the well) decreases. To restore pressure within older fields, petroleum geologists have learned to inject CO_2 into the geological formation that holds the oil or gas. The resulting increase in reservoir pressure improves oil and gas production rates. This practice is often cited in discussions of sequestration, but in the petroleum business CO_2 injection is done with little concern for whether the CO_2 remains permanently in the ground. The CO_2 might make its way back to the surface in a few years or a few decades, or it may remain in the ground indefinitely. No one knows because most fields are not monitored for leakage.

The best experiments with respect to injection of CO_2 into the ground for the purposes of sequestration have been carried out by Norway, which, since the 1990s, has injected modest but still significant amounts of CO_2 deep beneath the floor of the North Sea. These sites are carefully monitored and their performance has been the subject of detailed computer simulations. Available data indicate that the CO_2 will remain in the formation indefinitely.

But while Norway's experiments work well for Norway, the situation is more complicated than this brief description suggests. Sequestration is a local technology. Large amounts of CO_2 are produced in Tokyo and Cleveland, for example, but the geological conditions near these two cities differ from each other and from those beneath the North Sea. Consequently, if power producers in these areas decide to sequester CO_2, they will have to tailor their approaches to their locales just as Norway did. Sequestration methods are dictated by local conditions.

(continues)

(continued)

Given the local nature of sequestration, questions about cost, the local geology, safety (a large leak could be deadly), and the length of time that the CO_2 would remain sequestered are all local questions with local solutions. Each site must, therefore, be evaluated on its own merits, which increases costs and implementation time. If an affordable and permanent local sequestration site does not exist, then the CO_2 must be emitted, or transported by pipeline—a very expensive solution—to a suitable sequestration site. A great deal of work remains to be done to make this potentially important technology a reality.

EMISSIONS CONTROLS

Because global warming is so much in the news, it is easy to concentrate on CO_2 emissions, which are a major contributing factor to climate change, and ignore other emissions. To do this, however, would be to misunderstand the emission-control problems faced by power-plant operators today. Coal-fired power plant operators are not, as a matter of law, required to control CO_2 emissions. Even if they wanted to do so, they could not. The necessary technology and infrastructure are not in place. But in many nations, coal-fired plant operators are legally required to control sulfur dioxide (SO_2) emissions. This is not an easy job.

Sulfur dioxide, which is a product of the combustion of coal, is a key ingredient to smog. It creates respiratory problems among those living downwind of the plant and contributes to acid rain. Reducing SO_2 is a vital part of limiting environmental damage due to the operation of coal-fired power plants.

Coal plant in Colstrip, Montana. Coal-fired power plants are seldom located near homes of the well-to-do. *(D. Hanson and Northern Plains Resource Council)*

Reducing SO_2 emissions begins with the choice of coal supplier. In the United States, coal deposits with the lowest average sulfur content tend to be located in the West. But the chemical makeup of coal varies in many different ways, and power producers do not simply order from the supplier with the lowest levels of sulfur. Western coal, for example, also has a lower heating value than many Midwestern and Eastern coals, so more of it must be burned to produce the same amount of thermal energy. But even after comparing the sulfur content and heating values of Western coal deposits to those of coal obtained from Midwestern and Eastern mines, many producers find that it still makes economic sense to burn Western coal, because the expense associated with SO_2 control makes Midwestern and Eastern coal economically unattractive.

Some power producers use other strategies with respect to procuring fuel. Some plant operators prefer to mix higher sulfur, higher heating value coals with lower sulfur, lower heating value coals to obtain a fuel with moderate levels of sulfur and a higher heating value than can be obtained from low-sulfur, low–heating value coals alone. Blending coal in this way makes the fuel clean enough so that the plant can satisfy emissions requirements.

Another factor affecting the choice of supplier is transportation costs: For those power plants located in the East or located adjacent to a large supply of higher sulfur coal, it may make good economic and even environmental sense to burn higher sulfur coal.

If an operator decides to use a higher sulfur coal, it is generally washed in a process described in chapter 4. Washing does not remove all of the sulfur—even in theory—but it can remove a substantial amount. Much of the sulfur present in high-sulfur coal was deposited during the time that the plants that comprise the coal were still growing. More sulfur may have been deposited within the coal beds by groundwater with a high sulfur content that flowed through the coal beds after they were formed. (See chapter 2 for a more complete description.) Much of the sulfur that was deposited in either of these ways can be washed from the coal because it is mixed *with* the coal but is not tightly bound *to* the coal. But some of the sulfur present in coal is part of the coal matrix in the sense that it was part of the plant material that formed the coal. This sulfur cannot be separated by the washing process. After coal is washed, therefore, it will contain some depositional sulfur—this was missed during the washing process—as well as the sulfur that is an integral part of the coal matrix.

When the coal is finally pulverized and burned in the furnace, the sulfur remaining in the coal powder is released. Because much of the sulfur was removed during the washing process, it might seem that the amount of SO_2 remaining would be too small to matter, but a large coal-fired power plant will burn in the neighborhood of 20,000 short tons of coal (18,000 metric tons) *each day*. Even if only 0.5 percent of the weight of the coal is sulfur—and this would be considered low-sulfur coal—the plant would still produce 100 short tons (90 metric tons) of SO_2 each day. This is the problem: So much coal is used each day that even small amounts of pollutants—and small is measured as a percentage of the weight of one unit of coal—can create large problems. The solution is to install a so-called flue gas desulfurizer (FGD), also called a scrubber.

There are many different scrubber designs, and engineering firms remain hard at work trying to create new scrubbers that are more effective and cheaper to install and use. Scrubber technology is mechanically and chemically complex. The scrubber is inserted between the combustion chamber and the smokestack. Scrubbers are efficient in the sense that they remove most of the targeted pollutants from the hot combustion gases. Whatever is left after passing through the scrubber—water vapor, CO_2, and the pollutants that escaped—goes up the smokestack and into the broader environment.

Flue gas desulfurizers work by injecting material into the combustion gases that reacts with the SO_2 and makes removal of the product possible. There are three basic methods. One is to inject a liquid in the form of a fine aerosol. Dissolved within the liquid is a material, often lime, that reacts with the SO_2. The SO_2 is absorbed by the droplets. The droplet/combustion gas mixture is then passed through a mist eliminator. The mist eliminator removes the droplets that absorbed the SO_2 before passing the cleaned gases along. This type of FGD, called a wet scrubber, is very common. Other scrubbers inject a fine dry material—again lime is a common choice—which adsorbs the SO_2. The dry particles are then removed prior to emitting the combustion gases to the broader environment. The third type is a semidry scrubber, which suspends the reacting material in a mist that evaporates in the combustion gases, leaving a dry adsorbing material suspended in the combustion gases.

These are then removed to prevent them from entering the broader environment.

The amount of SO_2 removed by a scrubber is substantial, but a very small percentage of the SO_2 produced during the combustion process still finds its way up the chimney. From a legal perspective, the operation is considered a success as long as emissions fall below the maximum amount allowed by law, but additional incentives have been created in the form of SO_2 emissions trading (see the sidebar "Carbon Trading Markets") to encourage power producers to further reduce their emissions. The technology and legislation used to control SO_2 emissions have reduced emissions to a tiny fraction of what they once were. These efforts, both technological and legislative, to reduce SO_2 emissions are often cited as important examples of environmental successes.

Alternative Uses for Coal

Coal can be used to produce gaseous fuels and liquid fuels. Processes by which such conversions can be accomplished have been known for a very long time. When implemented on a commercial scale, these technologies require large amounts of energy and large amounts of coal. Historically, they have also been associated with large amounts of pollution. In fact, in the past, nations have employed these technologies only when they felt that they had no other choice. Despite past problems, interest in more efficient and less environmentally disruptive conversion processes is growing rapidly. This chapter describes some of the ways that coal has been used to manufacture liquid and gaseous fuels.

METHANE FROM COAL

The first successful attempts to convert coal into a gaseous fuel, called coal gas (or sometimes town gas), were made by the British inventor

The Great Plains Synfuel Plant. Each day it converts approximately 18,000 short tons (16,000 metric tons) of lignite into methane and in the process produces a number of commercially valuable coproducts. *(Basic Electric Power Cooperative)*

William Murdock (1754–1839). He demonstrated the idea in 1792 by using the coal gas that he produced to illuminate his house. In 1812, the Gas Light and Coke Company began laying gas pipes along London streets to supply gas for street illumination. In 1816, Baltimore became the first city in the United States to install a coal gas infrastructure, initially for street illumination. Soon cities throughout Europe and the United States were manufacturing their own coal gas.

Natural gas was not an option at the time. The technology for drilling natural gas was primitive, and the technology for building large diameter pipes to transport natural gas from the fields where it could be produced to the distant cities where it was needed did not yet exist. By contrast, coal was easy to transport because it could be brought to market by wagon, barge, ship, or train. A large city would have several gasworks, each with its own service area, manufacturing gas that was distributed to the street lamps by small diameter pipe. (It was not until the second half of the 19th century that coal gas became a common fuel for cooking and heating as well as illumination.) The switch from gas illumination to incandescent lighting began shortly after the American inventor Thomas Edison (1847–1931) invented the lightbulb. The transition was slow, however, because technical innovations in the gaslight industry made gaslights competitive with incandescent lights for decades and coal gas remained competitive with natural gas even longer. The coal gas industry finally entered a period of sharp decline during World War II when innovations in pipeline technology made natural gas both inexpensive and widely available.

The process by which coal gas was produced was simple and dirty. Coal was heated in a low-oxygen environment. The heating produced carbon monoxide gas, hydrogen gas, methane, tars, sulfur dioxide gas, and other materials. The gas mixture produced by the heating was cleaned as well as the technology of the time allowed. The result was a gas that consisted primarily, but not entirely, of (poisonous) carbon monoxide, gaseous hydrocarbons, and hydrogen gas, all of which are flammable. The mixture was distributed by pipeline to consumers. Town gas had a relatively low heating value—lower, for example, than natural gas—but town gas was an important innovation. It could be distributed via pipeline from a centralized location; it burned cleaner than coal and was, when it was introduced, the best fuel available for street illumination.

Much has changed since the last U.S. gasworks went out of business early in the 1960s. Coal can now be used to manufacture methane, to produce a gaseous fuel that is essentially identical with natural gas. It has always been more expensive to manufacture synthetic natural gas than to obtain natural gas from a well. The United States government has, nonetheless, spent billions of dollars developing and deploying the technology. The initial motivation for the project was the oil crises of the 1970s (see chapter 8). The most significant outcome of the federal effort to produce methane from coal is the Great Plains Synfuels Plant in Beulah, North Dakota.

Constructed near a huge lignite field, the Great Plains Synfuels Plant consumes six million short tons (5.4 million metric tons) of coal each year to produce 54 billion standard cubic feet (1.5 billion m^3) of synthetic natural gas. (A standard cubic foot is the amount of natural gas occupying one cubic foot [0.028 m^3] at 60°F [16°C] at one atmosphere pressure.) A variety of other commercially valuable co-products are manufactured in the process, including the fertilizers ammonium sulfate and anhydrous ammonia, phenol, which is used in the manufacture of plywood, and carbon dioxide, which is used to enhance recovery rates at nearby Canadian oil fields. (Carbon dioxide injection into oil fields is described in "More on CO$_2$ Sequestration" in chapter 5.) The process by which this so-called syngas is manufactured is complex and energy intensive, but essentially it involves taking apart the molecules that comprise the coal and reassembling them together with hydrogen to form methane molecules (CH$_4$) plus byproducts. (Commercial natural gas is almost pure methane.)

The Great Plains plant had numerous problems, technical and managerial, during the early years of its operation. The plant operators have, however, continued to improve the operation of the plant and decrease the cost of syngas production. In the meantime, the cost of natural gas has continued to increase. Today, domestic natural gas production is no longer adequate to meet domestic demand, and imports are rising. The day may soon come when the Great Plains plant becomes economically self-sufficient. Meanwhile it continues to serve as a huge laboratory for syngas technology.

COAL AS A TRANSPORTATION FUEL

The technology for the production of transportation fuels from coal is currently founded on the pioneering work of German chemists

Franz Fischer (1877–1947) and Hans Tropsch (1889–1935), who, in 1925, discovered what is now known as Fischer-Tropsch synthesis. Broadly speaking, the process for turning coal into liquid fuels involves two basic steps. First, the coal is gasified to produce something similar to town gas. The next step, which was Fischer and Tropsch's contribution, is to liquefy the gas produced in step one. The resulting liquid can be further refined to produce diesel fuel, aviation fuel, gasoline, and other hydrocarbons.

Initially, Fischer and Tropsch's discovery was interesting but too expensive to implement on a commercial scale. At the time it was cheaper just to buy oil and refine it. But during World War II, Germany, which has few domestic petroleum resources but large domestic supplies of coal, was cut off from imported petroleum and found it necessary to use coal to meet its need for liquid transportation fuels. After the war, there was little reason for Germany to continue to use coal to produce transportation fuels because the price of oil was so low. Later, faced with international sanctions, South Africa used the Fischer-Tropsch method to convert some of its coal reserves into liquid transportation fuels.

The Fischer-Tropsch process can, in theory, be used to supplant some of the United States' petroleum imports as well. The United States imports about twice as much oil as it produces domestically. The heavy dependence of the United States on imported oil has serious consequences for the environment, the economic health, and the energy security of the nation. With its huge supplies of coal, the United States could supplement domestic production of oil by turning coal into a liquid fuel, but the cost, economic and environmental, for coal-to-liquid technologies is quite high. In the United States, the elected officials with the most enthusiasm for this project tend to represent states with large coal industries.

But if the costs of coal-to-liquid technologies are too high for the United States, they may not be too high for China, which is currently developing commercial-scale coal-to-liquid production

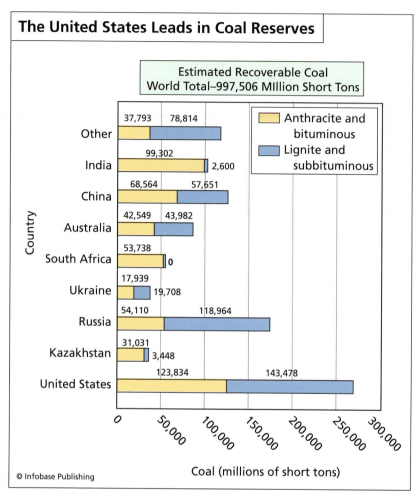

The United States Leads in Coal Reserves

Estimated Recoverable Coal
World Total–997,506 Million Short Tons

Legend:
- Anthracite and bituminous
- Lignite and subbituminous

Other: 37,793 | 78,814
India: 99,302 | 2,600
China: 68,564 | 57,651
Australia: 42,549 | 43,982
South Africa: 53,738 | 0
Ukraine: 17,939 | 19,708
Russia: 54,110 | 118,964
Kazakhstan: 31,031 | 3,448
United States: 123,834 | 143,478

Y-axis: Country
X-axis: Coal (millions of short tons) — 0, 50,000, 100,000, 150,000, 200,000, 250,000, 300,000

© Infobase Publishing

U.S. coal reserves are much larger than those of any other nation. *(EIA)*

facilities. Progress has been uneven, but China seems determined to displace a significant fraction of its oil imports with this technology regardless of the environmental costs involved.

A SOURCE FOR HYDROGEN GAS

Hydrogen is often described as the fuel of the future. There are two reasons: First, hydrogen burns easily and produces a great deal of

Engineering the Climate

Modern societies depend upon the consumption of enormous quantities of fossil fuels, and this is unlikely to change any time soon. For economic reasons, rapid change is almost impossible. Trillions of dollars have already been spent on coal-fired power plants, natural-gas–fired power plants, airplanes, automobiles, trucks, trains, furnaces, etc., and the infrastructure necessary to support them. There is not enough money on the planet to replace all of this in the short term. Moreover, even as the efficiency of combustion technologies improves, atmospheric CO_2 levels continue to increase because of rapid growth in countries such as India and China, once minor sources of CO_2 emissions.

One method of stabilizing the global climate that does not depend upon reducing the rate at which CO_2 is added to the atmosphere is to reduce the amount of the Sun's energy reaching Earth's surface. Keep in mind, the principal deleterious effect of a high level of atmospheric CO_2 is that it causes Earth to retain more of the Sun's energy. Consequently, one method of stabilizing the climate is to reduce the amount of solar energy reaching Earth's surface. With this approach, CO_2 levels could be allowed to continue to increase with few ill effects.

There are several methods for reducing the amount of solar energy reaching Earth's surface. One method that has received attention involves distributing material in the middle and upper stratosphere. (The stratosphere is that section of the atmosphere that begins between 4 and 11 miles [6–17 km] above Earth's surface and extends to an altitude of about 30 miles [50 km]. The altitude of the lower boundary of the stratosphere depends on the latitude at which it is measured; it is higher near the equator than near the poles.) One approach to shielding the atmosphere from the Sun's rays would be to add minute particles of sulfur to the stratosphere. About 770,000 short tons (700,000 metric tons) would be necessary, an amount that is small compared to the billions of tons of CO_2 that are emitted into the atmosphere each year.

The idea is not new. Scientists have long known that large volcanic eruptions sometimes reduce Earth's temperature for a few years even

(continues)

(continued)

though the "insertion" of volcanic material into the stratosphere is not done in a way that optimizes its effects on global temperature. Done correctly, most of the sulfur placed in the stratosphere would remain aloft for several years. Eventually, of course, the sulfur particles would settle out, and the supply of stratospheric material would have to be replenished. The projected costs associated with this technology are tiny compared to the cost of replacing the world's fossil fuel infrastructure—even assuming that such a massive overhaul of the world's energy infrastructure could be accomplished on short order, which it cannot.

Opinions on the wisdom of this approach vary. In an April 2008 paper in the journal *Science,* a group of researchers cast doubt about the wisdom of using sulfur, which in addition to lowering global temperatures might also deplete the ozone layer. The lead researcher, Simone Tilmes of the National Oceanic and Atmospheric Administration, in an interview with the BBC indicated that ozone destruction did not necessarily mean that sulfur should not be used: "If you have to make decisions, you need to know what is good about it and what is bad about it. With this scheme the bad side is definitely the ozone depletion, but you can cool the climate."

To be sure, there are technical and environmental difficulties associated with lofting hundreds of thousands of tons of sulfur or some other material high into the atmosphere. But perhaps an even more difficult problem would involve forming a consensus among the world's governments about what the optimum climate should be. If humanity can adjust Earth's temperature, who, in effect, gets to set the thermostat? The question is not trivial. There will always be winners and losers whenever the climate changes. If the climate is to change by design, will those disadvantaged by the plan agree to its implementation?

heat in the process. (Measured on a per unit mass basis, burning hydrogen releases three times as much thermal energy as burning the same amount of gasoline.) The second reason many support the use of hydrogen as a fuel is that during the process of combustion, two hydrogen molecules combine with one oxygen molecule to produce two water molecules plus heat, according to this chemical equation:

$$2H_2 + O_2 \rightarrow 2H_2O + heat$$

Hydrogen gas consists of hydrogen molecules, each of which is composed of two hydrogen atoms chemically bound together. As previously mentioned, two oxygen atoms will also combine to create a two-atom oxygen molecule. In the equation, H is the chemical symbol for a hydrogen atom and O is the symbol for an oxygen atom; the subscript "2" indicates that two atoms are bound together to form two-atom molecules. The symbol H_2O on the right side of the arrow is the chemical formula for the water molecule.

Hydrogen is, therefore, remarkably clean burning, but it is far from an ideal fuel. Hydrogen gas does not exist freely on Earth. While there are large quantities of hydrogen on Earth, almost all of these hydrogen atoms are bound to other atoms—many are bound to oxygen atoms in the form of water molecules, and many are bound to carbon atoms to form hydrocarbons. Consequently, one cannot prospect for hydrogen gas the way that one prospects for natural gas. Instead, one must manufacture it. To accomplish this, one needs a source of energy and a source of hydrogen atoms, called a feedstock. Water, of course, can be used as a feedstock. Electricity can be used to split the water molecules and produce hydrogen and oxygen gases in a process called electrolysis that is familiar to most high-school students. But electrolysis is expensive in the sense that more energy is consumed (in the form of electricity) than is recovered by burning the resulting hydrogen or by using the hydrogen to power a fuel cell. And what is true for electrolysis is true generally: It takes more energy to make hydrogen fuel than one recovers from its use.

Today, the cheapest, most widely used methods for producing hydrogen depend upon fossil fuels as feedstock. The only fossil fuel cheap enough and abundant enough to satisfy a large hydrogen market is coal, which could serve as both the energy source to power the process and the feedstock for the production of hydrogen.

To be sure, other methods of producing large amounts of hydrogen exist. Specially designed high-temperature nuclear plants could, for example, produce enormous amounts of hydrogen from water. The technology for such plants already exists, but there would have to be a huge demand for hydrogen before any investor would consider building such an expensive project. Solar energy could be converted into electrical energy, which could be used in electrolysis, but currently the contribution of solar energy to the nation's grid is practically negligible. It is not clear that without substantial improvements in solar technology that solar could meet large-scale demand for affordable hydrogen, even if the demand for hydrogen already existed.

In addition to difficulties involved in producing hydrogen, there are also unsolved problems with storing and transporting it efficiently. Nevertheless, the knowledge that supplies of natural gas and oil will continue to be tight for the foreseeable future continues to spur efforts to find ways to use hydrogen in place of these primary energy sources for heating, power generation, and as a transportation fuel. If these research efforts are successful, demand for hydrogen will certainly soar. A future hydrogen economy may well be based on coal.

Price versus Cost and National Coal Policies

The demand for coal depends on a complex mix of factors: the price of coal, the costs of producing coal and the way that these costs are distributed, government policies toward coal, coal's relationship to other energy sources, and even simple financial inertia. (Power plants are expensive to build and are generally financed with long-term loans; operators sign long-term contracts to buy fuel and to provide power, and, properly maintained, coal-fired power plants have long lives. All these factors make a plant difficult to shut down once it has begun operation.) This chapter begins by examining how coal-fired power plants are used by power producers and some factors affecting the price of the electricity that they produce. The chapter concludes with some remarks on the national policies of three major coal-consuming nations.

Coal train. A tremendous infrastructure has developed to support the coal economy in the United States. *(Ponderosa Ranch)*

PRICING COAL

Politicians and economists enjoy speaking about "the" free market, but there are many free markets. The conditions under which each market operates are determined by a complex set of laws and regulations. Some economists assert that fewer laws and regulations are better, but this view has not always been supported by the available data. The situation with respect to coal is complex.

In the United States, electricity producers constitute almost the entire market for coal. Electricity production can be divided into two functions: base load power production and peak power production. The distinction is important. Hospitals, around-the-clock manufacturing concerns, and other similar users all require power

24 hours per day, seven days per week. Their consumption patterns are highly predictable. These users establish a "floor," or minimum level of demand that must be satisfied no matter the weather or any other variable factor that might influence demand. The power required to meet this minimum demand is called base load power. Base load power producers must be highly reliable, and power plants used to supply base load power must be able to operate for prolonged periods of time without interruption. In the United States utilities purchase base load power through long-term contracts.

But the total demand for electricity can fluctuate far above the base load requirement. Schools, stores, banks, and other businesses all consume more power during the day than during the evening. The amount of power that these institutions require depends, in part, on the time of day, the day of the week, and the weather. Air-conditioning, for example, is energy-intensive, and only short-term predictions about temperature can be made reliably. These time-dependent additional demands are called the peak power requirements of the system. (Sometimes demand is more finely divided, but for purposes of this book all demand is classified as peak or base load.) Peak power is purchased on an as-needed basis—sometimes the day before it is needed, sometimes a few hours before and sometimes just minutes before. Peak load power is only furnished for relatively brief periods of time, so peak power generating stations must be relatively easy to start and stop.

Currently, coal and nuclear power produce most base load power. They are reliable, relatively inexpensive to operate, and capable of operating for long periods without interruption. Natural gas, which can be started with little preparation, can be used to provide either base load or peak load power, but the ever-escalating price of natural gas means that it is increasingly used for peak power, which tends to be sold at a premium. (Regions that depend upon natural-gas–fired plants for base load power tend to have high electric rates.) These three sources—coal, nuclear, and natural gas—constitute most of the power-generating capacity of the United States and most other large economies. (Hydroelectric power, which provides less than 10 percent of the total power output in the United States, will not increase because all of the most productive sites for hydroelectric facilities in the United States have already been developed. Future hydroelectric development is limited in many other nations for the same reason. Other renewables—wind, solar, and geothermal, for example—currently produce only a tiny fraction of the total power, either base or peak, and for a variety of reasons this situation is unlikely to change, at least over the medium term.)

Coal has, therefore, an important and narrowly defined place in the power sector, where, within its niche, it has few competitors. A coal-fired power plant's profitability is determined in large measure by the cost of fuel, which is currently inexpensive. But the reasons that coal is inexpensive are complex and revealing.

Economists sometimes make a distinction between the words *price* and *cost*. The price of an item is what appears on the bill; the cost is interpreted more broadly. In this sense, the costs of an item can be measured, in part, by the effects that the production of the item has on the workers that produce it, by any subsidies that the producer receives, and by the effects that the producer in question has on the community in which it is located; costs can also be measured in terms of the effects that production has on the broader environment.

By contrast, business people often have a narrower view of the meaning of the word *cost*. In particular, from the point of view of coal mine operators, costs are what they must pay to operate. Profit, measured on a per-ton-of-coal basis, is the difference between the price that they charge power producers for the coal that they sell them and the money that they pay to bring the coal to market. Similarly, for power producers, costs are what they pay to produce a unit of power. Profit is the difference between the price that they charge and the costs that they pay.

As with most businesses, both mine operators and power producers seek to maximize the difference between the price they charge and the costs they pay. One method of increasing profits is to increase efficiency. For mine operators, this might mean increasing the number of tons of coal produced per miner per hour. For power producers, this might mean increasing the number of kilowatts produced per unit of coal burned. Statistics of all sorts show that in the United States, for example, both mine operators and power producers have become increasingly efficient.

But there are other ways to increase profits without raising prices. Cost shifting, the transfer of costs (however interpreted) onto other, often unwilling, individuals or institutions, is just as effective a way to increase profits as is increasing efficiencies.

With respect to mine operators, many of the operators' costs of doing business have historically been borne by miners. In some cases they still are. Shifted costs may appear in the form of unnecessary occupational hazards. Poorly ventilated mines, inadequate safety training, and inadequate safety equipment are forms of cost shifting. These costs are surely borne by the miners. Additional profits resulting from these shifts, if any additional profits result, are enjoyed by the investors. Similarly, communities situated near coal mines may bear the brunt of cost shifting in the form of serious water pollution problems resulting from decisions by mine operators to forgo environmentally sound but more expensive mining practices. These types of cost-shifting behaviors may enable mine operators to keep the price of coal low, thereby enabling them to maintain or increase profits without increasing the price of their product.

Another example of cost shifting involves power-plant emissions, especially CO_2 emissions. Currently, power producers that depend on coal (as well as all other users of fossil fuels) emit all of the CO_2 they produce directly into the atmosphere. This effectively distributes the cost of coal consumption among all of the inhabitants of the planet because everyone pays costs associated with climate change.

Power producers, however, keep all of the profits generated by power production. Costs are shared, profits are not. If those power producers with significant CO_2 emissions paid the costs associated with emitting CO_2, coal-fired power plants would be much less attractive investments relative to the available alternatives. Similar statements can be made of other types of power-plant emissions.

Some attempts have been made to require businesses that emit CO_2 to pay for the privilege. Implementing such a system has proved difficult. First, there is no generally agreed upon formula for assigning a monetary value to CO_2 emissions. How much should one pay for emitting a ton of CO_2 into the atmosphere? Second, cheap electricity is a public good. Nations that are most dependent on coal-fired power plants—the United States, China, and India, in particular—have the most to lose by increasing the price of electricity generated by coal-burning power plants. Some other nations can act more aggressively to cut CO_2 emissions with little penalty and little effect. Norway, for example, generates most of its electricity with hydroelectric plants; it has already instituted a carbon tax.

Finding ways to assess the true costs associated with mining and burning coal are imperative if coal is to be priced in a way that reflects the sum of its societal and environmental costs. This is not to say that coal should not be used to generate electricity. But without an accurate assessment of the true costs of each fuel and each type of power generation technology, it is difficult to determine which fuels and technologies should be developed and how that development should proceed.

NATIONAL ENERGY POLICIES

The following brief descriptions of national policies with respect to the use of coal-fired plants reveal the different ways that coal technology is perceived and developed. All three of the nations, the United States, Germany, and China, have significant domestic supplies of coal, and all three are heavily dependent on coal-fired power plants.

Government policies are important in determining the ways that coal reserves are developed. *(Sean Linehan, NOAA, NGS, Remote Sensing; courtesy, NOAA)*

In the United States about half of all electric power is produced by coal-fired plants. Somewhat more than 90 percent of all coal consumed in the United States is consumed for the purpose of generating electricity. Three very large domestic industries depend on the continued use of coal-fired power plants: the coal mining industry, the power generation industry, and the rail transport business. (Essentially all coal in the United States is moved either by rail or by rail in conjunction with some other form of transport, such as barges.) A fourth group with a strong interest in maintaining inexpensive coal are the many individuals and businesses that benefit from relatively inexpensive electricity. Taken together these groups constitute a powerful pro-coal lobby. Policy changes that might increase the cost of coal or re-

Carbon Trading Markets
"How Much Is Air Worth?"

The easy answer is that air is worth a lot. Many people would say that air is priceless, but their actions belie their words. Virtually everyone engages in discretionary activities—vacation travel, for example—that contribute to CO_2 emissions. The same general pattern of discretionary pollution applies to many businesses. In fact, historically most individuals and companies have behaved as if the atmosphere had very little value. Experience has shown that resources that are not valued are overused.

One possible approach to assigning value to air is to charge access to the atmosphere—at least for certain activities. In this case, what is restricted is the *total amount* of a particular pollutant that can be added to the atmosphere each year. By enforcing such a restriction, the atmosphere becomes (from an economic point of view) a finite resource. Finite resources have value; infinite or unlimited resources do not. It is *because* air is shared by all and without limit, this theory claims, that air is valued by no one.

These ideas have already helped to motivate the sulfur dioxide (SO_2) emissions-trading system that was developed in the United States after passage of the 1990 Clean Air Act. Sulfur dioxide is a powerful pollutant, and in 1990, U.S. fossil fuel plants, mostly coal-burning plants, were emitting 18 million short tons (16 million metric tons) of SO_2 per year into the atmosphere, a tremendous environmental load. The goal of the emissions trading system was, over time, to reduce emissions to 9 million short tons (8 million metric tons). The method was to allocate so-called "emissions allowances." Each emissions allowance was good for a particular amount of SO_2—that is, the holder of the allowance was permitted to emit as much, but no more, SO_2 as was permitted by his allowance. (Because the number of emissions allowances was limited, so was the total amount of SO_2 that could be emitted. The limit created the value for the allowance.) If it was necessary to emit more SO_2—for example, a power producer might need

duce demand for coal have historically been difficult to institute over the objections of those groups most heavily dependent on cheap coal.

to burn more coal than anticipated and as a consequence emit more SO_2 than it was allocated—the power producer could purchase additional allowances on the open market. If the power producer was able to meet demand without emitting all of the SO_2 it had been allocated, it could sell the remainder. The buyer and the seller only had to agree on a price. This was the price that had to be paid to pollute in excess of the allowance. The method by which SO_2 emissions were measured was determined by the government.

The trading scheme forced each polluter to create an SO_2 budget. Cutting emissions could, in theory, pay for itself because the emissions credits could be sold for a profit to others who were less successful at staying within their budget.

The program worked. There has been no evidence of cheating and plenty of evidence that SO_2 emissions decreased in a way that has largely justified the confidence of the plan's backers. To be sure, the system was not without difficulties. Initially, some of the dirtiest industries just bought whatever credits they needed to cover their shortfalls. Over time, however, they also found ways to make reductions in emissions.

This type of program is now being advocated as a method for reducing CO_2 emissions. A carbon emissions trading market has, in fact, already been established in Europe, although too many credits were issued at the outset, and a surplus of credits, not surprisingly, made each individual credit worthless.

There are enough differences between SO_2 and CO_2 that a CO_2 trading scheme has thus far proven difficult to establish despite the experience with SO_2 emissions trading. This is not to say that efforts aimed at creating a CO_2 emissions-trading scheme must fail, but only to point out that they have not yet succeeded.

The United States government subsidizes the coal industry in a number of ways. The subsidy of most interest here is sponsorship of research into so-called clean coal technology. Billions of dollars have been spent by the Department of Energy to search for ways to minimize the environmental impacts associated with coal consumption without making coal consumption uncompetitive relative to alternative energy sources. If successful, clean coal technology would enable those sectors of the economy that depend upon the consumption of enormous quantities of inexpensive coal to continue to enjoy the benefits of the coal economy. The research, which is always described as aimed at improving the environment, also provides a competitive advantage to coal-dependent industries and individuals at little direct cost to them.

There are other interesting, if unintended, consequences to United States government policy. Although the United States is working hard to develop coal plants that would burn coal far more cleanly than the standard pulverized-coal plant, its environmental laws have had the effect of keeping old coal-fired plants in operation. Over time the United States' environmental regulations have become more stringent, leaving coal-fired power producers to choose among the following options: (1) continue to operate an existing coal plant, (2) build a new coal plant, (3) build a natural gas–fired plant, or (4) convert a coal-fired plant to one that burns natural gas. But because older—and, in general, dirtier—coal-fired power plants are exempt from certain of the newer, more stringent regulations, power producers have found it more economical to keep older, less efficient plants in operation rather than to build new ones. Consequently, the average age of U.S. coal-fired power plants is increasing rather than decreasing. The situation is not expected to begin to change substantively until 2015, when the rising price of natural gas is expected to force power producers to build new coal-fired plants, rather than the natural gas–fired plants that they currently prefer to build to supplement output from the older coal-fired plants.

In Germany, coal is the nation's only significant domestic source of fossil fuel. Roughly 50 percent of German electrical generating capacity is produced by burning coal. Despite its reliance on coal, German electricity rates are some of the highest in the European Union.

As with the United States, the German government generously supports advanced research into so-called clean coal technology, but the resemblance between the two national policies ends there. Over the short-term the German government is committed to the rapid replacement of the nation's older and less efficient plants with new, more efficient ones. This is accomplished using a CO_2 emissions law that will take effect in 2012. After 2012, operators of coal-fired plants have three options: first, they may adopt sequestration technology; second, they may acquire emissions allowances by supporting "climate projects" outside of Germany; or third, they may participate in a CO_2 emissions-trading scheme. But if a plant operator brings a modern coal-fired plant on line before 2012, it is guaranteed emissions rights for its first 14 years of operation.

More generally, the German approach to coal-fired power plants is shaped by four more general policies. First, for environmental reasons, the German government is committed to minimizing the nation's reliance on coal. Consequently, government policies are aimed at minimizing the growth of coal-fired power generation. Second, Germany has instituted fairly strict conservation measures and so electricity demand has increased more slowly than in, for example, the United States. There is, therefore, no need to expand the supply of electricity as fast in Germany as in the United States. Third, Germany is seeking to expand that sector of its power production capacity that is often described as renewable. There are, for example, more wind turbines in Germany than in any other country in the world. Finally, Germany has undertaken a program to close all its nuclear power plants, a policy that assures a heavy reliance on coal for the indefinite future. Taken together, the mix of energy sources by which Germany powers its economy is changing, but its reliance on coal is not. New coal-fired power plants, which will continue to operate for decades, are currently being brought online. These plants, although modern in design, are not equipped with sequestration technology. As with other nations with large domestic coal supplies, the German electricity sector remains committed to coal-fired power technology, and it will remain a significant source of CO_2 emissions for the foreseeable future.

With respect to China, now the world's biggest consumer of coal, the EIA predicts that its economy will grow at an average rate of 6.5 percent per year until 2030. Much of this growth will be powered by coal. In China almost one-half of the coal consumed is used by the manufacturing sector, especially in the production of iron and steel. Both electricity and steel production are carried out with scant regard for the environmental implications of unrestricted coal consumption.

Although China's power production infrastructure is new, the technology on which it depends is not. Keeping pace with its surging demand for electricity will mean adding numerous additional coal-fired power plants over the next two decades. Given China's preference for less expensive, less efficient technologies, increases in power plant emissions of all sorts will continue to rise in proportion to generating capacity. It is expected that coal consumption will double in the power-generation sector and in the manufacturing sector between 2004 and 2030. Without changes in plant technology, emissions can be expected to double as well. In the absence of modern coal-burning technology, coal consumption of this magnitude will present a significant environmental hazard. It is not clear how much Chinese power producers are willing to pay for cleaner coal-burning technology. Presumably, as the costs associated with the use of the less efficient technologies become more apparent, the costs of newer, more expensive, and cleaner technologies will appear to be more of a bargain. But this is by no means certain. And no matter how attitudes change in the future, it is important to keep in mind that once a plant is built, there is a tendency, for economic reasons, to keep it running as long as possible.

The environmental challenges with respect to coal consumption are numerous, and it is not clear that they will all be successfully met. Nevertheless, because inexpensive electricity is a public good, because coal is the most abundant of all fossil fuels, and because there are so few alternatives, coal will almost certainly prove at least as important an energy source during the first half of the 21st century as it was throughout the 20th century.

PART 2

Oil

A Brief History of Oil

Oil is the essential fuel. In the transportation sector, there are no good alternatives to oil. An oil alternative would have to be produced in bulk—the United States, for example, consumes more than 20 million barrels of oil *each day,* mostly for transportation— and any oil alternative would have to be safe to transport and store. Ideally, it would also burn at least as cleanly as oil. Additionally, today's technology would have to be able to make use of it. There is no ready alternative to the jet engine, for example. Oil is currently the only fuel that meets all of these criteria, and this is not surprising because the world's trillion-dollar transportation infrastructure was built to run on oil. While it is true that biofuels such as *ethanol* and *biodiesel* have made some inroads into the transportation fuels market, there is not enough agricultural land on the planet to produce sufficient biofuels to displace more than a small fraction of the gasoline and diesel currently in use. And even the most optimistic scenarios do not foresee hydrogen displacing oil during the next two decades: Assuming that technical problems associated with the production, distribution, and storage of hydrogen are solved during the next decade, the infrastructure required to support a hydrogen economy would still take many years to construct. Shale oil and coal oil are expensive to produce, and the processes involved in their manufacture are difficult to justify from an environmental viewpoint. Finally, cars and trucks that obtain all of their energy from batteries will remain a niche market until revolutionary changes in battery performance are effected.

Meeting worldwide demand for oil is one of the great business, security, and technological challenges of today. This chapter outlines some of the most important milestones in the history of oil as a commodity.

THE OIL BUSINESS PRIOR TO 1950

The oil business began in 1859 when the first commercial oil well was struck in Titusville, Pennsylvania, but this was not the first time that people had seen oil—not even in Titusville. Many people, living in different areas of the country, were already familiar with oil because some oil deposits were close enough to the surface that oil would sometimes seep onto ground, and drillers in search of water would sometimes strike oil by accident. Initially, however, oil was perceived as having little value. Eventually it was discovered that, properly processed, oil could be used as a clean-burning illuminant. This was an important observation because other fuels used for lighting were already recognized as unsatisfactory: Tallow candles produced lots of smoke, and whale oil, the illuminant of choice, had become increasingly expensive as whales became harder to find due to unrestricted whaling. Prior to 1859, coal oil, a liquid illuminant that could be extracted from coal, was being manufactured even in New Bedford, Massachusetts, the center of the whaling industry, but quantities were limited. The time was right for oil.

Signal Hill oil field, California. Early oil fields left most of the oil in the ground even as they maximized environmental damage above. *(The Bancroft Library, University of California, Berkeley)*

After overcoming numerous technical and financial hurdles, the American Edwin Laurentine Drake (1819–80) drilled Titusville's first successful commercial oil well. His discovery quickly led to the world's first oil boom. The oil field that Drake had uncovered was large and rich, and although his first well was not especially productive when measured in barrels produced, it was, at least, profitable. Problems soon arose, however, when later wells began to gush oil. In 1861, Pennsylvania's famous Empire well began to produce oil. Initially, oil flowed out of the Empire well at a rate of 3,000 barrels per day. This was unprecedented, and the well operators were unable to stop the flow. At the time, oil produced from a well was stored in barrels, but the Empire produced so much oil that the well's owners soon exhausted the region's supply of barrels. Oil poured onto the ground. The well owners built earthen dams in an attempt to capture their unanticipated riches, but the oil overwhelmed their dams. Again in 1861, even while the Empire well was still flowing at a rate of 2,500 barrels per day, another well, the Phillips well, began to produce at a rate of 4,000 barrels per day. Taken together these two wells provided enough oil to meet the entire world's demand for oil at that time. But many other wells had also been brought into production, and consequently, supply far exceeded demand. In January 1860, oil sold for $19.25 per barrel; by January 1861 the price had dropped to $1.00 per barrel, and by January 1862 the price had dropped to 10 cents per barrel. By January 1865 the price had climbed back to $8.25 per barrel. These wild price fluctuations made planning impossible for producers and consumers alike.

No matter the price they were offered for their product, producers maintained production at the maximum possible rate. Once a successful well was struck, the well owner would erect a second derrick immediately adjacent to the first. If the second was successful, a third would be erected. Early oil fields looked like forests with oil derricks in place of trees. Not to be outdone, competitors would acquire drilling rights on properties immediately adjacent to producing properties and begin to line the boundary of their property with wells of their own in an attempt to take as much oil as possible from the producing property. All early producers operated under what became known as "the rule of capture," a legal doctrine that stated that whatever oil could be extracted from a given well belonged to the owner of that well no matter the origin of the oil. The Pennsylvania Supreme Court eventually gave formal voice to the idea, which, whatever its legal merits, proved to have disastrous economic and environmental consequences.

The effect of the rule of capture was to force each operator to remove as much of "their" oil as possible before they lost it to competing wells. But the effect of sinking many wells in close proximity to each other in order to remove oil as fast as possible also meant that the oil field pressure, the force that caused oil to flow up through a well without the use of a pump, dropped quickly. In fact, fields would "dry up" with more than 80 percent of the oil left in the ground. Operators simply abandoned their wells and moved to new locations. Left open, the abandoned wells further reduced the reservoir pressure as pressurized natural gas, which is frequently found together with oil, escaped. It was a learning process, and as producers ruined one oil field after another, they incidentally discovered some of the physics of petroleum. As early as 1866, some were counseling against the wasteful procedures then in use, but to no avail. It was law, not physics, that determined the economics of recovery, and the economics were that no matter the consequences, once a well began producing oil, that oil needed to be extracted as fast as possible.

Because producers could not leave the oil in the ground in order to produce oil at a more efficient rate and in a more efficient manner, they stored it above ground: barrels, wooden tanks, iron tanks, and sometimes just huge pits dug into the ground were filled with oil. Spills and leaks were common. The landscape was soaked in oil.

For the first few decades of the 20th century, oil recovery methods were still determined by the rule of capture. Large new discoveries caused enormous price fluctuations and tremendous waste. Producers felt compelled to extract oil without regard to the effects their efforts had on the price of oil or the physical condition of the oil reservoir itself. The Spindletop discovery, made outside of Beaumont, Texas, in 1901, is a case in point. The first producing well practically exploded with oil. The well gushed 70,000 to 100,000 barrels per day for the first 20 days before drilling crews were able to cap it. Additional wells were quickly added, and the price of oil plummeted. It finally reached three cents a barrel, less than the cost of a glass of water in the oil field. Undeterred, producers placed derricks so close together that sometimes the legs of the neighboring derricks touched.

Standard Oil

In 1863, the American industrialist John D. Rockefeller (1839–1937) and two partners established an oil *refinery,* a business for converting crude oil into commercially valuable products, near Cleveland, Ohio. It was a profitable venture. In 1870, Rockefeller, together with associates Maurice B. Clark and Henry M. Flagler, formed Standard Oil, which soon became one of the largest and most influential companies in history. They already owned a large refinery business by this time, but they recognized that purchasing the competition and controlling the railroads, not better refinery technology, were the keys to controlling the oil markets. Standard Oil entered into agreements with the railroads that stipulated that Standard Oil would receive large rebates for shipping oil with them, rebates that were not offered to Standard Oil's competitors. This drove down the cost of business for Standard Oil relative to the cost of business of its competitors.

Soon, using its immense wealth and market position, Standard Oil made it increasingly difficult for its competitors to ship oil at any price, and when its competitors could not meet their contractual obligations Standard Oil bought them out. By 1877, Standard Oil controlled not only the shipment of oil by rail out of the Pennsylvania oil fields but also the pipelines that connected individual wells to the storage facilities at the railroad stations. Flush with cash, Standard Oil also bought many refineries and merged with many other refining companies in more distant mar-

In 1915, production of gasoline exceeded that of kerosene for the first time. Early refineries had considered gasoline more of a safety hazard than a product of value. They had concentrated on the manufacture of kerosene for its value as an illuminant. But with the success of the internal combustion engine, gasoline became the main product of the refinery process. (A refinery is a facility where crude oil is converted into commercially valuable products.) Oil had become a transportation fuel and a heating fuel rather than a source of illumination.

kets. By 1880, more than 90 percent of all refining was done by Standard Oil. It owned most of a rapidly evolving and strategic industry, and what it did not own, it controlled.

Many of Standard Oil's tactics were deliberately crafted to be anticompetitive. In response to efforts by states to rein in Standard Oil's monopolistic practices, the company reconstituted itself as a so-called trust, a complex and opaque administrative structure that made state and federal oversight and regulation difficult. Unable to control the trust, the government sought to dismantle it. In 1911, after many years of litigation, the federal government finally forced the breakup of Standard Oil. In some sense, this occurred almost too late to matter. Most of the originally aggrieved parties had either already sold out to Standard Oil or been driven from business. In any case, during the intervening years, Standard Oil's dominance had begun to fade. The effects of the breakup were, however, still significant. Some of the 33 smaller companies that resulted from the breakup of Standard Oil were Standard Oil of New Jersey (later Exxon), Socony (later Mobil)—Exxon and Mobil have since merged to form Exxon Mobil—Standard Oil of California (later Chevron), and Atlantic Richfield, Standard Oil of Ohio (Sohio), and Amoco, all three of which are now part of BP.

By 1920, petroleum engineers and geologists could correctly estimate that then-standard recovery techniques were leaving at least 80 percent of the oil stranded in the ground—that is, the fields became nonproductive when no more than 20 percent of the oil had been removed. These observations were beginning to have an effect on the governments of oil-producing states, but progress remained slow.

In addition to oil, natural gas was also wasted in enormous quantities. Natural gas is often found under high pressure in the same formations as oil. It is, in part, the pressure exerted by the natural gas within the reservoir that causes the oil to flow up the pipe and out of the well. Removal of the natural gas causes a corresponding drop in reservoir pressure. Natural gas can, therefore, serve one of two functions: It can be extracted and sold as a fuel, or it can be left in the oil reservoir to maintain pressure and so aid in the recovery of the oil. But because there was so much natural gas, and because the infrastructure for transporting natural gas was limited (as were the early markets for this important natural resource), there was often no market for the enormous volumes of natural gas that rose out of wells. Moreover, although the technology existed to reduce the rate at which reservoir pressure diminished—even in the 1860s, compressed air was sometimes injected into wells to aid recovery of the oil—many producers failed to see the value of reinjecting the natural gas extracted from a well back into the reservoir. So the gas was vented directly into the atmosphere. It is estimated that from 1922 until 1934, U.S. oil fields vented 1.25 billion cubic feet (35 million m³) of natural gas directly into the atmosphere *each day*. This was about the same amount as was captured.

A great deal of the oil found its way into the atmosphere as well. Producers, anxious to capture as much oil as possible, often pumped enormous amounts of it into leaky aboveground storage tanks or even earthen pits, where the lighter, more valuable components of the crude oil, the components that would otherwise have been used to produce gasoline, for example, simply evaporated. In this way, from 10 to 25 percent of the value of the crude was lost to evaporation, but from the producers' point of view, it was not, at least, lost to the competition.

Oklahoma, which at the time produced more oil than any state other than Texas and more oil than Venezuela, the Soviet Union, and Mexico, the three most productive oil-producing nations after the United States, began to try to limit waste. Initial results were mixed because conservation attempts in Oklahoma simply gave a competitive advantage to producers in other states. There was a general recognition that the system under which oil was recovered was a wasteful one, but efforts to reform it met with little success until the Great Depression. Faced with declining demand and the enormous reserves of the recently discovered oil fields of East Texas, the oil-producing states turned to the federal government for help. The result was the Interstate Oil Compact Commission. Formed in 1935, the original members were Colorado, Illinois, Kansas, New Mexico, Oklahoma, and Texas. Other oil-producing states joined later.

Oklahoma governor Ernest Whitworth Marland (1874–1941) wanted a cartel-like organization that would impose broad restrictions on production in order to maintain prices, an idea adopted by OPEC, the Organization of Petroleum Exporting Countries, many years later. Texas governor James V. Allred (1899–1935) wanted the compact's authority restricted to those measures necessary to prevent the physical waste of oil and gas. Allred's goal was directly realized, but the measures needed to prevent waste entailed to some extent the regulation of production and other measures and these served to indirectly support the price of natural gas and oil, Marland's chief goal.

A pattern had developed: Oil producers and government regulators entered into an uneasy partnership from which both benefited. Oil and natural gas were to be produced domestically at rates that provided reasonable and reasonably stable profits for producers and simultaneously maximized recovery rates. But it was also during the late 1930s and 1940s that domestic production slowly fell behind domestic demand. In 1948, for the first time, the United States became a net importer of petroleum.

THE OIL BUSINESS AFTER 1950

In 1952, during the presidency of Harry S. Truman, the United States began legal action against the world's seven largest oil companies—

Oil rig in the Varg oil field, which is located in the North Sea. *(James Holt)*

Standard Oil of New Jersey, Standard Oil of New York, Standard Oil of California, Gulf Oil, Texaco, Royal Dutch Shell, and Anglo-Persian Oil Company—for conspiring to illegally fix the cost of oil. (Of the seven, four remain. Standard Oil of New Jersey became Exxon; Standard Oil of New York became Mobil, and Exxon and Mobil merged to become Exxon Mobil. Standard Oil Company of California became Chevron, and Chevron, Texaco, and most of Gulf Oil eventually merged to form today's Chevron; and after several large mergers, Anglo-Persian Oil is now BP.) Friction between the federal government and the large oil companies was not new but neither was cooperation. Price stabilization had actually been a goal of the United States government during the late 1930s and early 1940s. Harold Ickes, then U.S. Secretary of the Interior, had wanted to improve access to the huge reserves of Middle East oil without disrupting the domestic pricing of petroleum, and to accomplish this he requested the cooperation of the seven majors, as they were called. On the other hand, the profits enjoyed by these already enormously rich firms, which came, in part, from colluding to fix the price of foreign oil, could not help but inspire resentment in a nation that still remembered the abuses committed by Standard Oil, and by 1952, U.S. policy had swung from cooperation to confrontation. But the government was unable to find a solution to the anticompetitive practices of the majors. The problem was that an industry of enormous importance to the world was, in many ways, controlled by just seven companies, who, if they competed at all, managed to do so in ways that enriched all of them. They were multinational companies engaged in a complex, secretive, and vital business. They were, in some ways, the equal of the government that sought to constrain them.

The world's seven major oil companies also earned the enmity of the producing nations of the Middle East. Decades earlier, the countries of the Middle East had entered into "exploration and production agreements," also known as "concessions" with various combinations of the seven major companies. These were complicated agreements in which the producing countries sometimes traded important aspects of their sovereignty, including their ability to tax the oil companies, for modest payments by the oil companies. It was the companies, not the countries, that determined the size of the payments and the conditions under which they were made because at that time only the companies had the means to extract, ship, refine, and market the oil. Their client states, including those nations with the world's largest deposits of oil, had none of these capabilities. This situation was largely maintained, to the benefit of the seven major oil companies, throughout the 1950s and 1960s. Even the founding of the Organization of Petroleum Exporting Countries (OPEC) in 1960 had little immediate effect.

Rapidly growing post–World War II economies required ever-larger amounts of petroleum. Between 1950 and 1970, U.S. oil consumption more than doubled. Western European consumption increased by a factor of nine, and Japanese consumption increased by a factor of 100 as Japan became the second largest consumer of petroleum after the United States. During the first few years of the 1970s, oil consumption continued to increase and imports from the Middle East accounted for much of the growth. This was especially apparent in the United States, where imports were 37 percent higher in 1972 than they were in 1971 and another 31 percent higher at the beginning of 1973 than they were in 1972. This happened during the time that U.S. production had "peaked"—that is, domestic oil production had permanently stopped increasing.

On October 6, 1973, war broke out in the Middle East. Israeli forces fought against those of Syria and Egypt. Israel bombed oil terminals in Syria. These terminals had a combined capacity of 800,000 barrels per day. On October 16, OPEC suspended communications with oil companies and unilaterally announced oil price increases of about 70 percent. This was the end of the system by which the seven major international oil companies set the prices for oil. On October 19, as the 1973 war continued, the United States announced a substantial increase in military aid to Israel. The next day, in retaliation, Saudi Arabia placed an embargo on oil shipments to the United States. Other oil-producing nations in the region followed the Saudi example. A similar embargo was placed on the Netherlands for its support of Israel.

The war ended three weeks after it began. The embargos were maintained for several months, but they had limited effect on supply. The effects of the price increases on the world economy were, however, profound, and they were just beginning. In December 1973, at an OPEC meeting, the Shah of Iran announced another major price hike. He gave two reasons: First, he asserted that the price of oil should reflect the cost of developing alternative sources of energy, and second, he asserted that Iran had to industrialize before its oil supplies were exhausted, and the increased oil income would make this goal possible. Oil prices were now roughly three times higher than they were prior to the 1973 war, and the OPEC nations were the principal beneficiaries.

The major consuming nations responded to the price hikes by instituting conservation measures, developing alternative supplies of (non-OPEC) oil, increasing research into energy alternatives, and where possible, switching from oil to other fuels such as coal, natural gas, and nuclear energy. The United States, for example, authorized a $10-billion program for energy research, and it sped the development of an oil pipeline connecting the oil fields on Alaska's North Slope with the south Alaskan port of Valdez. By contrast with the United States, France responded to the threat of much higher oil prices by aggressively building an electricity generating system that depended largely on nuclear energy. (Today, France continues to operate the most comprehensive and advanced system of nuclear reactors in the world.) Change took time, but oil-consuming nations did respond.

Surpluses in the world supply of oil began to appear as early as 1974; surpluses caused, in part, by an economic slowdown due to the spike in oil prices. Over the next few years, prices for oil stagnated, and coupled with inflation, partly caused by the 1973 spike, the real value of a barrel of oil slowly decreased. Efforts to develop non-OPEC sources of oil caused OPEC's share of the world oil market to diminish throughout the 1970s.

In 1979, prices increased sharply again, but for entirely different reasons. This increase was triggered by events in Iran, a major oil-exporting nation. The shah, deeply unpopular with many Iranians, was driven into exile in January 1979. During the months immediately preceding and following his departure, there was a great deal of civil unrest in Iran. In particular, all Iranian oil sales were halted for a few months. As a consequence, oil markets were thrown into disarray. Prices climbed sharply, but they also experienced substantial fluctuations. Later in 1979, as prices began to settle, Iraq and Iran entered into a long and bloody war, and once again prices climbed. Most OPEC nations reaped enormous profits during this time, not because of clever or disingenuous business practices but because of investor uncertainty about supplies from an unstable region. Here are the results: In 1978, OPEC oil revenues were $136 billion; by 1980, OPEC revenues had climbed to $287 billion. Iran's and Iraq's misfortunes were a windfall for other oil producers.

Tripling the price of oil, which is what happened during the second oil crisis, again depressed world economic growth. This second crisis further spurred efforts already underway to switch fuels. In the United States, in 1973, for example, 18 percent of all electricity was generated by burning oil; that same year in Japan, 64 percent of all electricity was generated by burning oil. By 1984, the United States was only generating 5 percent of its electricity with oil, and in Japan the figure was down to 34 percent.

Oil markets were changing rapidly. Although OPEC nations could, through coordinated action, force up the price of oil by limiting production, they could not force anyone or any nation to purchase OPEC oil. OPEC income in 1982 was $207 billion, down $80 billion from 1980, and the volume of oil that OPEC was selling had diminished by 47 percent as countries and companies sought to diversify their sources of supply. In order to maintain high prices, OPEC had to continue to cut back on supply, but because prices had stagnated, supply cutbacks translated into lost revenue. The task of maintaining the price of oil by restricting production was left to OPEC's most powerful member, the world's largest oil producer, Saudi Arabia. During the first quarter of 1979, Saudi Arabia produced at a rate of 9.5 million barrels per day, but by August of 1985, Saudi Arabia was producing oil at a rate of 2.2 million barrels per day. Almost half this amount was for domestic use. Even the Saudi government could not indefinitely maintain such drastic limits on production. The price of oil soon collapsed as Saudi Arabia resumed more normal levels of production.

During the early 1980s, OPEC members met regularly to decide upon a price structure for selling oil, but not all of the oil that these nations sold was sold at the agreed upon price. Initially, relatively small amounts of oil were sold on the "spot" market. (*Spot transactions* are similar in concept to the transactions that occur at filling stations when consumers purchase gasoline.) Spot purchases occurred when, outside of the OPEC price structure, a buyer and a seller arrived at a mutually satisfactory price for a one-time shipment of crude oil of an agreed-upon grade for immediate delivery. As the 1980s progressed, spot purchases became increasingly common, and the amount of oil purchased in this manner continued to increase. Meanwhile OPEC nations worked to maintain a base price for their product, but with greater awareness of their limits to do so.

Today, oil markets remain volatile, in part, because supplies are very tight and because a number of the most important oil producers are in areas that are politically and militarily unstable. Small changes in demand or in production can, therefore, result in large changes in price. Instability makes long-range planning difficult for all parties. But even in politically more stable regions, the interests of oil-producing countries are sometimes inimical to those of oil-consuming countries and vice versa. There is a natural tension between the two groups. Nor are the sources of tension purely economic. Historically, oil-consuming countries have not hesitated to contribute to the oppression of citizens in some producing nations—even to the extent of using military force—in order to secure oil. Policy makers in consuming nations remember 1973 when some producing nations sought to use oil as a weapon and in the process caused economic disruption throughout the world.

Despite the suspicion, and sometimes hostility, that exists on both sides, producing and consuming nations are, in an economic sense, almost completely dependent on each other. Many of the main oil producers have little income other than from the sale of oil. The hope that the Shah of Iran expressed in 1979, for example, that Iran would industrialize before its oil supplies were exhausted, has not yet come to fruition. The situation is the same in many other producing nations. Meanwhile, there is still no practical substitute for petroleum as a transportation fuel, and consuming nations require vast quantities of the fuel in order to run their economies. Producing and consuming nations are, then, locked into their often uncomfortable business relationships for the foreseeable future.

The Geology and Chemistry of Oil

One of the central questions for those interested in energy, and especially for those interested in oil, is, "How much oil is there?" This much is certain: Oil is a finite resource, and currently more than 70 million barrels of it are being withdrawn from the ground each day. Under these circumstances, world production must eventually peak and then decline. To appreciate some of the constraints on the amount of oil in the world, one must understand something of the geology of oil, and that is the first goal of this chapter.

With respect to oil, chemistry is just as important as geology. There are as many types of crude oil as there are oil reservoirs. Each type of oil has its own characteristic chemical and physical properties: Some crude oils flow easily; some do not; some contain more sulfur, a harmful pollutant, than others; some types of oil are richer in the molecules that comprise gasoline, and some are richer in the molecules that comprise asphalt. The second goal of this chapter is to examine some of the chemical and physical characteristics of crude oils and how they affect its value.

OIL AND GEOLOGY

Oil begins with the bodies of single-celled aquatic organisms. The ancestors of today's blue-green algae, plankton, and the very delicate, highly symmetrical diatoms, for example, are all thought to have contributed to the formation of today's oil. Other determining factors are the chemistry of the water in which these organisms lived and died and their relative numbers. All oil deposits require millions of years to form. They are, therefore, from the point of view of humanity, irreplaceable.

The oil-forming process begins when the bodies of these minute and ancient creatures are entombed in clays or other very fine sediments. The formation of this matrix of organic matter and clay is the first step in the formation of oil. Protected by the clay, the organic matter is slowly transformed into a material called kerogen. Meanwhile, the twin process of erosion and deposition continue to rework the landscape above the kerogen deposits, changing the distance to the surface of the clay-kerogen matrix. If erosion predominates and the kerogen is exposed to the atmosphere, no oil is formed, but if deposition predominates, the kerogen will slowly be buried deeper and deeper. Pressures and temperatures increase with depth, and if the process is continued long enough, the kerogen will experience temperatures and pressures conducive to the formation of oil and natural gas. (Oil and gas formation are generally thought to occur in the region between 2,500 and 16,000 feet [760–4,900 m] beneath Earth's surface. Much deeper and only natural gas will be produced.) These burial and conversion processes require millions of years, and the details of the entire process—the pressures and temperatures actually experienced by the kerogen, for example—further affect the quality of the oil produced.

Tiny droplets of oil (and trace amounts of natural gas) interspersed with fine-grained sediments cannot be extracted economically, which is another way of saying that no matter how much oil is contained in such a deposit, it is not an oil reservoir. But sometimes pressure differences will cause the petroleum (a word that is sometimes used to refer to a mixture of oil and natural gas) to begin to flow. The petroleum does not flow alone; it is usually mixed with water. The petroleum-water mixture is often unable to flow upward because of impermeable layers of sedimentary rock, such as sandstone or shale that, millions of years earlier, had formed above it. Consequently, the liquid creeps laterally beneath a ceiling of rock until it is expelled from the fine-grained sediments into coarser grained, more permeable rock.

The great sheets of impermeable rock that lie above the oil and help to channel its flow are usually not horizontal. Subjected to slow-acting but powerful geological forces, they are often folded; sometimes they are broken. From the point of view of oil prospectors, the folds are the most important. These folds may sometimes appear on topographical maps as huge, very gently rising domes called anticlines. The petroleum will follow the rising surface of the anticline and become trapped within the dome structure, and it is within this petroleum "trap" that the oil begins to accumulate. The accumulation process takes eons, but because the petroleum cannot escape upward through the barrier and cannot escape downward once it has entered the down-turned bowl-like structure of the anticline, the amount of trapped petroleum can only increase. Given enough time, tremendous amounts of petroleum can accumulate inside a large anticline, and while other types of "petroleum traps" exist, for a long time the search for oil was largely confined to a search for anticlines, which, experience has shown, are the richest repositories for oil.

Once inside the anticline, buoyancy forces cause the fluids to separate. Oil, which is lighter than water, will float on top of the

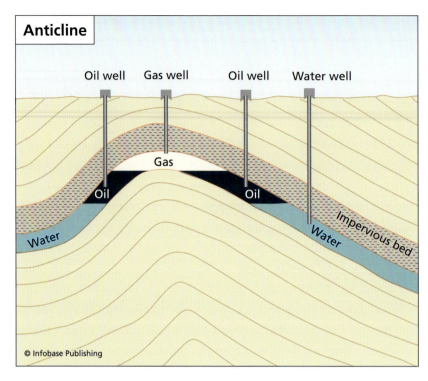

Anticline

Oil well   Gas well   Oil well   Water well

Gas

Oil   Oil

Water   Water   Impervious bed

© Infobase Publishing

These formations called anticlines have historically held many of the richest deposits of petroleum.

water layer, and methane, which is the principal component of natural gas, will often separate from the oil and form a "gas cap," a bubble of methane that forms above the oil and within the pores of the rock that lies beneath the dome-like barrier of the anticline. (Sometimes, if the pressure inside the deposit is high enough, some or even most of the methane will remain dissolved within the liquid oil. It will bubble out as the oil flows up a well in a way that is similar to the way that soda will bubble when the container is opened.)

From this description it is clear that oil formation is a haphazard process. Many things can disrupt the process, but if all conditions are met, the results can be remarkable. The first supergiant oil field to be discovered—a supergiant is sometimes defined as any reservoir that holds upwards of five billion barrels of recoverable oil—is in East Texas. Discovered in 1930, it reached peak production in 1972 when operators extracted an average of 212,000 barrels per day. It has already produced more than five billion barrels, and operators continue to draw off a little more than ten thousand barrels of oil every day. As large as the East Texas deposit was, it is dwarfed by the world's largest oil field, the Ghawar field, which is located in Saudi Arabia. Discovered in 1948, the Ghawar field contains an estimated 123 billion barrels of recoverable oil. Roughly half has already been extracted. As might be expected, supergiant fields are difficult to hide, and it is believed that most, or perhaps all of them, have already been discovered. (A few, for example, may exist beneath the ice of Antarctica.)

Oil exploration is a continual process. Production and consumption are in a precarious balance. As much oil must be discovered as is produced in order to maintain that balance. As technologies evolve the demand for oil will change. The hope is that the oil will last as long as it is needed.

THE CHEMISTRY OF OIL

Crude oil, the term for oil as it is found in the ground, consists largely of carbon atoms and hydrogen atoms, 82–87 percent carbon and 12–15 percent hydrogen when measured by weight. Sulfur is often the third most common element in crude oil, but the amount of sulfur present varies greatly from sample to sample. Sulfur is a contaminant of crude oil just as it is a contaminant of coal, and during the refining process the sulfur is removed in order to reduce the pollution associated with combustion. Crude oil also contains small amounts of oxygen, nitrogen, sodium chloride (table salt), and various metals, and as with sulfur, the amounts of these materials vary from sample to sample.

(continued on page 122)

When Will the World Run Out of Oil?

This question sounds as if it is concerned with the physical capacity of Earth's oil deposits, but it is really more of a question about legislation, technology, and the price of oil, because the amount of oil that will be recovered depends on government regulations, the technology available to recover it, and the price for which the oil can be sold.

Oil companies do not always attempt to produce oil even when they are sure they have found it. Sometimes, for environmental or aesthetic reasons, a government may prevent a company from drilling. Even if a company is not prevented from drilling as a matter of law, it may still decide to refrain from drilling, at least for a while, for economic reasons. Oil companies are economic enterprises: Oil that is very expensive to produce may be too expensive to sell at a profit. Finally, oil companies may not have the technology to produce oil from a particular geological formation. A great deal of oil is always left in "depleted" oil fields, for example, but producing this oil may not be technically possible. An upper limit for the amount of oil that can be recovered under any circumstances is the amount of technically recoverable oil. This is the amount of oil used when estimating peak oil, that point in time after which oil production can no longer be increased by any means.

There is a long history of attempts to estimate when global peak oil production will occur. In 1972, the United Nations estimated that oil production would peak around 2000. They were wrong, as was Shell Oil when, in 1979, it estimated that peak production would occur in 2004. In 1998, the International Energy Agency (IEA), an agency of the United Nations, estimated that the peak in production would occur in 2014, but in 2000 the IEA weakened its estimate, predicting that peak oil would occur only after 2020. In 2003, Shell revised its forecast, predicting that peak oil would occur only after 2025. Perhaps the most thoroughly researched and widely quoted estimate was published in 2000 by the Energy Information Agency (EIA). The EIA's prediction—it actually produced a set of nine predictions—considered three different rates of economic growth, a high (3 percent) growth rate, a medium (2 percent) growth rate, and a low (1 percent) growth rate, and it assumed that oil consumption rises with the rate of economic growth. The EIA used three USGS estimates for the amount of technically recoverable oil worldwide. (Multiple USGS estimates reflect uncertainty about the amount of technically recoverable oil.) The USGS estimated that there are between 2.248 and 3.896 trillion barrels of technically recoverable oil. The EIA used this information to identify three different possibilities with respect to oil resource levels: the low oil scenario (2.248 trillion barrels), the medium oil scenario (3.003 trillion barrels), and the high oil scenario (3.896 trillion barrels). Given, therefore, an economic

(continues)

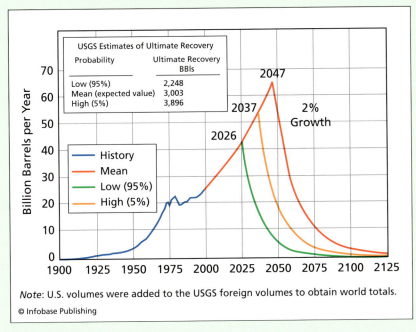

Note: U.S. volumes were added to the USGS foreign volumes to obtain world totals.

© Infobase Publishing

All mathematical models predict that once peak oil is reached, production falls off rapidly.

(continued)

growth rate and an estimate for the amount of oil remaining, and, finally, given some conservative estimates about the pace with which any alternative (nonoil) technologies will be introduced, the EIA forecasted the timing of peak oil for each of nine scenarios: the low economic growth rate/low oil scenario, which can be shortened to (low, low), and the other eight scenarios, which can be abbreviated as (medium, low), (high, low), (low, medium), (medium, medium), and so forth. The EIA's forecasts for peak oil given a 2 percent growth rate and the low, medium, and high estimates of technically recoverable oil are summarized in the graph on page 121.

(continued from page 119)

The carbon and hydrogen atoms found in crude oil are almost always found in the form of molecules called hydrocarbons, which consist of various combinations and configurations of carbon and hydrogen atoms. Some hydrocarbon molecules contain only a few carbon and hydrogen atoms; others contain many atoms. Some hydrocarbons are long, straight chains of carbon and hydrogen atoms, and others have more complex geometries. This variation is important because different hydrocarbons have different chemical and physical properties. Methane, for example, consists of one carbon atom bound to four hydrogen atoms. (Its chemical formula is CH_4.) It is the simplest of the hydrocarbons and at ordinary temperatures and atmospheric pressure, methane is a gas. There are other hydrocarbons that are liquid when the crude oil is first recovered, but when exposed to open air quickly evaporate.

The hydrocarbons found in crude oils are classified as belonging to one of three "series." The first series, called the paraffin series,

The petrochemical industry has generated important new products for more than a century in labs such as the one shown above. *(Alfa Laval)*

consists of molecules having the chemical formula C_nH_{2n+2}, where n is the number of carbon atoms in the molecule under consideration. Notice that when $n = 1$, this is the formula for methane given in the preceding paragraph, so methane, the principal component of natural gas, is a member of the paraffin series. More generally, if an element belongs to the paraffin series and it has n carbon atoms, it must contain $2n + 2$ hydrogen atoms. When n is less than 5, the elements of the paraffin series are gaseous at ordinary temperatures and pressures. Elements of the paraffin series with $5 \leq n \leq 15$ are liquids, and if n is greater than 15, materials composed of such molecules are very sticky liquids or solids. Molecules belonging to the paraffin series are the most common of all hydrocarbons in crude oil.

The second series, the elements of which are a little less abundant in crude oil than those of the paraffin series, is the naphthene series. These molecules have the general form C_nH_{2n}—that is, each molecule in this series has exactly twice as many hydrogen atoms as carbon atoms. An example of an element in the naphthene series is ethylene. Its formula is C_2H_4. Ethylene is often used to speed the ripening of bananas.

The third series, the elements of which form rings, is called the aromatic series. Hydrocarbons belonging to this series are, when compared to those of the paraffin and naphthene series, generally less common in crude oil. The most common hydrocarbon in this series is benzene, which has many industrial applications. The use of benzene in gasoline, once common, is now restricted for environmental reasons; it is carcinogenic. Hydrocarbons in the aromatic series have the form C_nH_{2n-6}. (Benzene's formula is C_6H_6.)

For the oil producer, the chemistry of crude oil first makes its presence felt when attempting to recover the oil. The chemical composition of crude oil affects its ability to flow. Some crude oils flow easily, and consequently are easier to recover. These are the so-called light and intermediate weight crudes. Saudi Arabia has large reserves of light crude oil, and light crude oil comprises about two-thirds of its production. So-called heavy crude oils flow more slowly and require more effort and expense to extract. Venezuela has large reserves of heavy crude oil. Some oil-like deposits of hydrocarbons, most notably the *oil sands* deposits located in northeast Alberta, Canada, do not flow at all. (See the sidebar "The Oil

Sands of Alberta" in the next chapter.) Oil sands are composed of hydrocarbons; they burn when ignited, and properly processed, oil can be recovered from them. Currently, they are often recovered via surface mining techniques.

If one thinks of crude oil as a mixture of gasoline, kerosene, lubricating oil, fuel oil, asphalt, and other products, then refining, at its simplest, involves separating, or unmixing, the different components, also called fractions, of the crude oil. To separate the fractions, refiners make use of the fact that different hydrocarbons have different boiling and condensation points. This was the observation that allowed the first crude-oil refiners to separate kerosene and lubricating oil from crude oil. By heating the crude in a controlled way, they could make use of the fact that different fractions evaporated at different temperatures. And just as the fractions evaporate at different temperatures, the resulting vapors condense at different temperatures. By making use of these simple-sounding facts, refiners could "unmix the mixture." First, they heated the crude oil to create a rich chemical vapor. Then they separated out through condensation those fractions of economic value from those for which they had yet to discover a use. (Today refiners make use of everything in the crude oil, and they can do much more than simple distillation, which is the name of the process just described. The technology of refining is discussed in chapter 11.)

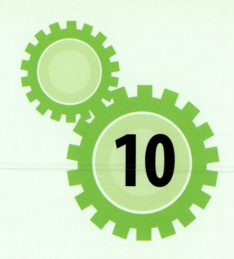

Meeting the
Demand for Oil

To appreciate how valuable oil has become, consider the amount of work and ingenuity involved in producing it. One of the goals of this chapter is to provide an overview of some of the technologies involved in obtaining oil.

But producing oil is only the first step. Oil is not distributed evenly throughout the world. Many of the world's leading oil-consuming nations have few oil resources of their own. Even the United States, one of the world's leading oil producers, consumes so much oil that it must import roughly two-thirds of its supply. Moving enormous volumes of oil from producers to consumers is another vibrant and technically challenging aspect of the oil business. Providing an overview of how oil is moved from one place to another is the second goal of this chapter.

The Trans-Alaska Pipeline System stretches from the northern settlement of Prudhoe Bay on the Arctic Ocean, across the Brooks, Alaska, and Chugach mountain ranges, to Valdez on Prince William Sound, a distance of about 800 miles (1,300 km). *(Southwest Research Institute)*

OBTAINING OIL

Today, a company in search of oil may well begin its search in the storage facility where it retains decades of data from previous attempts to find oil. Much of that data was acquired by sending small seismic waves into the ground and measuring how they moved through the earth. These waves were originally caused by small explosions. Later, other, more controlled methods were employed. The direction and speed of these waves are determined by the materials through which they pass. As might be imagined, it is far easier to cause the waves and even to measure the changes in them as they travel through the ground than it is to determine what those changes mean. As a consequence, earlier generations of petroleum engineers sometimes missed the telltale signs of oil in the data that they collected. But as better computers and better computer software have become available, the old data, which had already been collected, and so is, in a sense, free, can be reanalyzed. These techniques—collectively called "data mining"—played an important role in the discovery of the 1.45-billion barrel Agbami oil field in Nigeria.

In the field, oil companies still use seismic testing, but the tests and the accompanying analyses are becoming ever more sophisticated. With more powerful computers and better software, oil companies can now generate detailed three-dimensional images of an oil field, images that make more accurate decision-making possible. (In the 1950s, approximately one of every five wells drilled actually struck oil. Today, that figure is approaching one in two.) At some point in the process, the companies involved in the exploration of a particular area must decide whether or not to drill. (Exploring for oil is often a joint effort as companies attempt to share costs and distribute financial risks.)

Sophisticated technologies enable drilling crews to drill faster and more accurately than ever before. By way of example, drilling crews can, if they choose, drill down and then turn at a 90 degree angle and drill horizontally. Moreover, drills are outfitted with sophisticated sensors that use the drill pipe itself as a local area network (LAN) in order to convey large amounts of drilling information quickly to the operators. Not surprisingly, the price of drilling wells has increased as well. If the operators strike oil, they may drill several other wells in order to better determine the size of the field they discovered. Finally, they begin to extract commercial volumes of oil in a way that maintains the reservoir pressure (the force that pushes the oil to the surface without intervention on the part of the operator) as long as possible. But, depending on local conditions, reservoir pressure may, even with the best technology, begin to fail when as little as 10 percent of the oil has been recovered.

The producer, having spent a great deal of money finding a field and drilling the wells, is, of course, reluctant to leave most of the remaining oil in the ground due to the failure of reservoir pressure. The next step is to use so-called secondary recovery techniques. Two common and closely related techniques involve injecting materials into the oil field with the goal of forcing the oil up the well. Water is sometimes used to achieve this result. Other times, natural gas, which is separated from the oil at the wellhead, is injected back into the oil field to maintain reservoir pressure. These ideas are not new. Producers in Pennsylvania's earliest oil fields tried using air to accomplish the same result and with some success.

Depending on the physical characteristics of the field, secondary recovery methods may enable the operator to recover an additional 5–30 percent of the oil originally present. Put another way: When secondary recovery methods begin to fail, from 60 to 85 percent of the oil originally in the field may still be in place. Typically, about two-thirds of the oil remains in the ground when secondary methods fail.

With the failure of secondary recovery techniques, the operator may employ so-called tertiary techniques, which attempt to further facilitate the oil's ability to flow to the well. A common tertiary technique involves injecting carbon dioxide (CO_2) into the oil field. The value of CO_2 injection is that it performs two functions at once: Injection of CO_2 raises the reservoir pressure, and CO_2 reduces the viscosity, or stickiness, of the oil, causing the oil to flow more freely. Carbon dioxide injection is sometimes described as "CO_2 sequestration," but the reason for injecting CO_2 into these formations is to recover oil, and it is not at all certain how long the CO_2 will remain in the oil fields. It would be a happy coincidence if this CO_2 remained in place indefinitely.

Sometimes, when oil does not flow easily, it can nevertheless still be forced to flow by heating it. One method of heating oil involves injecting large amounts of steam into the oil field. By injecting

The Oil Sands of Alberta

Extracting oil from the oil sands of Alberta, Canada, is difficult and expensive. The process requires a great deal of energy and produces a great deal of waste. No one would bother with such a technically difficult and environmentally challenging project except that the oil sands of Alberta constitute one of the world's largest supplies of fossil fuel energy, and oil prices are now high enough to justify the effort.

Alberta's oil sands cover 54,000 square miles (140,000 km²), and contain 1.7 trillion barrels of bitumen, the fossil fuel that comprises the oil sands. The Alberta oil sands make Canada second only to Saudi Arabia as the nation with the world's largest reserves of oil. But unlike Saudi Arabia, which contains huge reserves of easily-extracted petroleum, Alberta's principal oil resource has more in common with tar than oil. Bitumen, although rich in energy, will not flow unless heated or chemically modified. Large quantities of the substance are, however, located near the surface. Consequently, the oil sands are often mined rather than drilled. The soil covering the bitumen, which is called the overburden, is scraped away, and enormous machines dig into the deposit. Only about 10 percent of a shovelful of oil sands is (on average) actually bitumen—oil sands are primarily sand and clay—and processing techniques currently recover only about 75 percent of the surface-mined bitumen. As a consequence, it requires 2.2 short tons (2 metric tons) of oil sands to produce approximately one barrel of oil.

Deposits close enough to the surface to recover using conventional surface mining techniques constitute approximately 20 percent of the total deposit. Most bitumen is located too far below the surface to access by surface mining techniques. Increasingly, operators are trying to recover these deposits via "in situ" techniques, which are more like those used to recover heavy oil. One technique, called steam-assisted gravity drainage, works as follows: Two wells are drilled into the deposit, an upper and a lower well. The upper well is used to inject steam into the deposit. (The steam is sometimes mixed with a chemical to further

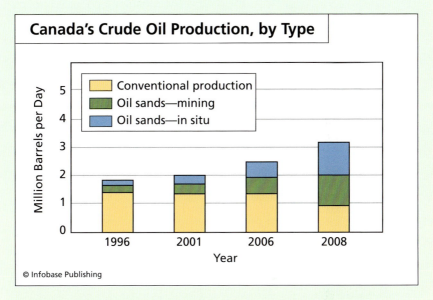

Canada's crude oil production by type. Canada and the United States, its main export market, are becoming increasingly reliant on oil produced from the tar sands. *(Canadian Association of Petroleum Producers; EIA)*

reduce the viscosity of the bitumen.) This forms a so-called steam chamber in which the bitumen is heated, causing it to flow, more or less, to the lower well, where it is extracted. At the surface, the water, sand, and other impurities are removed, and a "diluent," a material used to further reduce the viscosity of the bitumen, is added so that the bitumen can be transported via pipeline to a refinery. Because oil from the tar sands is expensive to produce, production rates have historically tracked the price of oil.

steam, the viscosity of the oil is reduced and the reservoir pressure is increased. The result is an increase in the rate at which the oil is recovered. In California, which has reserves of heavy high-viscosity oil estimated at about 40 billion barrels, steam injection is an important method of increasing oil recovery rates. Steam injection is also increasingly common in Venezuela, which has enormous reserves of heavy oil. A second method of recovering heavy oil involves burning some of the oil while it is still underground. As the oil burns it produces gases, raises reservoir pressure, and heats the surrounding oil, which (again) reduces viscosity, causing the oil to flow toward the well more easily.

Secondary and tertiary methods of oil recovery increase the cost of recovering each barrel. Oil extracted using these methods must be priced to reflect these higher costs. Despite the difficulties and costs, these sophisticated and expensive recovery techniques will play an increasingly important role as the production rates of older fields slow and discoveries of large new oil fields become less frequent. They remain of value as long as the price of oil is high enough to justify their expense.

TRANSPORTING OIL

The crude-oil market is the largest sector of international trade, whether that trade is measured in dollars or in the physical bulk of the material traded. One reason that crude oil is such a huge business is that consuming nations often have so little production capacity. The United States, which consumes upwards of approximately five billion barrels of oil per year, is an exception. It is both a major producer and a major consumer, but its consumption is so great that it must import roughly two-thirds of its supply. Some other major consuming nations, Japan and Germany, for example, import close to 100 percent of their crude oil.

Almost all oil is transported by one of two methods, pipeline or ship. Since the early days of the Pennsylvania oil boom, pro-

A supertanker fills up at the Al Basra Oil Terminal in the northern Persian Gulf.
(Jim Garamone; courtesy, American Forces Press Service)

ducers have attempted to use pipelines to transport oil, but for a long time they were hampered by inadequate technology. Early pipelines used small-diameter pipes because this was what the technology of the time allowed. Over time, engineers learned to manufacture sturdy large-diameter steel pipes and to join them in ways that were fast and reliable. There are a host of technical issues associated with these simple-sounding facts, and a pipeline infrastructure capable of moving large amounts of product took decades to create. Pipeline construction soared during World War II, as the United States sought to move large amounts of domestically produced crude and refined product without the use of tankers—then the preferred means of transport—because tankers were vulnerable to attack by submarines. A very extensive pipeline system now connects oil fields with refineries and refineries with distant markets. (Gulf Coast refineries, for example, are connected with the large markets located along the East Coast of the United States via pipeline.)

Pipelines are often well-suited for transporting large quantities of oil within the same country, and when countries have particularly close ties, as is the case with the United States and Canada, pipelines can also work well between them. Pipelines have proven less useful when transporting oil across unstable or unfriendly regions because pipeline flows are vulnerable to disruption. Pipelines are also poorly suited for bridging the distance across oceans.

The alternative to a pipeline is to use oil tankers. Many tankers, large and small, are in use today. Whether a producer chooses a large or a small tanker depends on several factors, including the route to be traveled and the type of cargo to be transported. The size of the ship matters, for example, when it must pass through a particular canal or strait or enter a shallow port. With respect to the type of cargo, refined petroleum products are generally transported in smaller ships, and crude is generally transported in larger ships. The further oil is transported, the higher the costs involved and the more likely it is that producers will choose a large tanker. Transportation costs reduce the profit margins of the producer, but the effects of these increased costs can be partially countered by choosing larger ships in order to benefit from economies of scale. All other things being equal, larger ships are simply more efficient carriers. This simple fact has motivated ship buyers to order the largest ships possible. Modern oil tankers are huge, with lengths routinely exceeding 1,000 feet (300 m). They are so large that they are usually loaded and unloaded from special platforms located far from shore. For this reason, most people outside of the business rarely see them up close.

Producers have discovered that the easiest way to maximize profits is to supply the needs of the energy markets nearest them. The largest supplier of foreign crude to the United States, for example, is Canada, and this oil is transported via pipeline. The next largest producers, all of whom currently supply roughly equal volumes of oil, are Mexico, Venezuela, Nigeria, and Saudi Arabia. Mexico and Venezuela are neighbors, of course, and Nigeria, because it is located on the west coast of Africa, also finds it profitable to send much of its oil to the United States. Saudi Arabia, by contrast, could probably make more profit by sending its crude oil elsewhere, but according to the EIA, Saudi Arabia tolerates the lower profit margins that long-distance transport to the United States entails for security reasons. In fact, most of the crude oil from the Middle East is sold in Asian markets, where higher profits can be made due to the shorter distances to those markets.

Some governments and companies have expended a great deal of effort making pipelines and tankers more spill-proof, and these efforts have had some success. Tanker spills and pipeline spills are discussed in chapter 13; here it is only necessary to note that the volumes of oil and oil products transported daily are enormous—tens of millions of barrels. Consequently, even when only a very tiny fraction of that total is spilled, the effect on the environment can be substantial. Spills are part of the price of the oil economy. Societies live with them for the same reasons that they live with so many other environmental effects associated with oil production and consumption: The world's transportation system runs almost exclusively on oil.

Creating and Consuming Petroleum Products

The production of transportation fuels is the primary business of most refineries, but other products are also important. This chapter provides an overview of the processes by which refineries convert petroleum into an array of products.

Fuels, of course, are created in order to be destroyed by burning. As gasoline, diesel, and jet fuels are burned, they produce heat and a complex mixture of products that are vented into the atmosphere. The production of heat is generally the sole purpose for which the fuels are created; everything else is a byproduct. As researchers have become more knowledgeable about controlling the combustion process, they have had more success controlling the chemical composition of the products of combustion. Better control enables engineers to further reduce the effects that the combustion gases have on the environment. The other goal of this chapter is to provide an overview of the combustion process.

REFINERIES

The refinery business is as old as the oil business itself. Unrefined crude oil is (and always has been) practically worthless. Refineries are easily recognizable as modern-looking, large, complex collections of towers and pipes illuminated with thousands of lights. But in the United States, most refineries are decades old.

As mentioned in chapter 9, crude oils are complex mixtures of hydrocarbons, molecules composed of hydrogen and carbon atoms. Hydrocarbons differ in size and in geometry. This variation in molecular properties and frequencies causes variation in the bulk properties of crude oils, such as their densities and their viscosities, and the amount of energy a sample of crude produces when burned. Because so many different substances are mixed together in crude

Oil refinery in Ireland. The transportation sector of virtually every nation is entirely dependent on petroleum. *(Europavalve)*

oil, the first task of the refinery is to decompose the crude oil into its various "fractions," the commercially valuable components of which it is composed. Essentially, the refiner unmixes the mixture. This begins, as mentioned in the discussion of the history of the oil business in chapter 9, with the process of distillation.

From the refiner's point of view, one of the most important differences among the many different liquids that compose crude oil is their boiling points. The boiling point of gasoline is different from the boiling point of kerosene is different from the boiling point of diesel and so on. The refinery operator heats the crude oil enough to evaporate all it—or at least almost all of it—and then passes the resulting vapor mixture, which is actually a mixture of the vapors of all the different fractions, through a long column called a fractional distillation column. The bottom of the column, is warmer than the top. As the vapor rises in the column it cools. As it cools, different fractions condense. A particular height within the distillation column, therefore, corresponds to a particular condensation point, so that at different heights different fractions can be drawn off.

Early refineries depended solely on distillation, but distillation was not entirely satisfactory even during the early days of the refinery business. The problem was that early refineries wanted kerosene, but after they had recovered the kerosene from the crude oil, they were left with less economically valuable fractions. These less valuable fractions, which from the refiners point of view were "waste," constituted a significant amount of the original barrel of oil. There was less demand for these materials, and so there was less profit associated with their sale. This was the refiners' dilemma: They had paid for the entire barrel but could only profit from part of it. During the 19th century and the early years of the 20th century, gasoline was part of the refiners' problem, a product for which there was little demand. The development of the automobile sector caused demand for gasoline to soar. In response, refiners sought to convert ever-larger portions of a barrel of oil into gasoline.

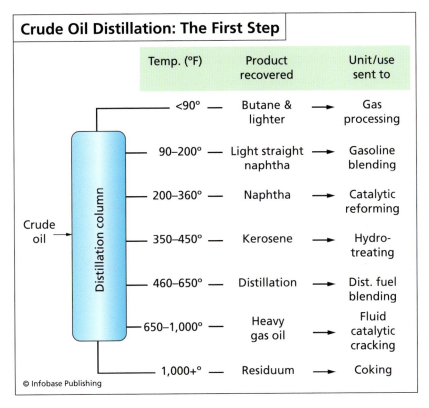

Crude Oil Distillation: The First Step

Temp. (°F)	Product recovered	Unit/use sent to
<90°	Butane & lighter	Gas processing
90–200°	Light straight naphtha	Gasoline blending
200–360°	Naphtha	Catalytic reforming
350–450°	Kerosene	Hydro-treating
460–650°	Distillation	Dist. fuel blending
650–1,000°	Heavy gas oil	Fluid catalytic cracking
1,000+°	Residuum	Coking

Crude oil → Distillation column

© Infobase Publishing

Distillation is the first step in the refining process.

The solution to the refiners' dilemma was the development of *cracking* technology, the chemical conversion of less desirable heavier hydrocarbon molecules into more desirable lighter ones. The idea behind cracking technology was understood by late 19th-century engineers and scientists, but as an industrial-scale process, cracking technology did not become widespread until the 1920s. Cracking technologies enabled refiners to increase the quantity of the lighter fractions that they obtained from each barrel of oil by converting some of the heavier hydrocarbons into lighter ones.

Not surprisingly, several different cracking technologies have been developed since the 1920s. Fluid catalytic cracking, for example, is a process that occurs at high temperatures and low pressures. The heavier fractions of the petroleum are mixed with a catalyst, a material that, while it is not consumed in the resulting chemical reaction, facilitates the conversion of heavier hydrocarbons into lighter ones. By contrast, a more expensive process, called hydrocracking, uses high temperatures and high pressures (greater than 130 atmospheres) to break the larger hydrocarbons into smaller, lighter molecules. In hydrocracking, hydrogen gas is introduced together with a catalyst. (The hydrogen gas is produced on site; refineries are some of the most important producers and consumers of hydrogen gas.)

Cracking technologies operate on a rich mixture of hydrocarbons. They reduce the average size of the hydrocarbon molecules on which they operate, but even after processing, the resulting mixture still contains some large hydrocarbons. In addition, some of the hydrocarbons produced by the cracking processes are too light—materials consisting of these molecules have boiling points that are too low—to have any value as gasoline. These lighter hydrocarbons are recovered and subjected to yet another process, called polymerization, by which they are combined to form larger gasoline-type hydrocarbons.

The cracking processes to which crude oil is subjected also have the effect of decreasing the density of the resulting product. One consequence of this density reduction is that an average 42 gallon (159 l) barrel of crude oil produces in excess of 44 gallons (167 l) of product, including (on average) 19.6 gallons (74.2 l) of gasoline, 10 gallons (37.8 l) of diesel and heating oil (they are essentially the same liquid), and four gallons (15 l) of jet fuel. These three transportation fuels, therefore, constitute roughly three-fourths of the entire output of the refinery.

THE CHEMISTRY OF COMBUSTION

Because gasoline consists of a number of hydrocarbon molecules, it is not possible to write down a single chemical equation to describe

Filling up with gas in the Netherlands *(Celia Barlow)*

what happens when gasoline is burned. There is no single "gasoline molecule." And there are important differences between what happens in carefully controlled laboratory conditions and practical automobile engines. Despite these complications, some useful statements can be made about the processes involved in burning gasoline.

Gasoline is burned in internal combustion engines, and the process begins by mixing hydrocarbon molecules with oxygen molecules. (Oxygen in the air is generally present in the form of two oxygen atoms chemically bound together to form an oxygen molecule.) By introducing energy into the fuel-air mixture in the form of a spark, a process is initiated in which the bonds that hold

⏻ The Refinery Business

Petroleum accounts for roughly 40 percent of all the energy consumed in the United States. Consumers spend in the neighborhood of $300 billion on petroleum products each year, somewhat more than $150 billion on gasoline alone. More than 90 percent of all the gasoline consumed in the United States is produced by U.S. refineries.

The technology used to refine petroleum is mature—that is, the ideas and basic technologies in use today were in use years ago—and when interviewed in 2002 for a study made by the RAND Corporation, industry executives stated that they did not anticipate that any radically new ideas would be introduced in the foreseeable future. That remains true today. It is not likely, therefore, that technological innovation will give one company a significant advantage over a competitor any time soon. How, then, do refineries compete?

In the absence of technological innovation, refineries could, of course, still compete on price, but this strategy has proved disastrous for the refiners. In 1981, there were 189 companies engaged in refining petroleum, and the industry utilized only 78 percent of its maximum capacity—that is, refineries were refining only about three-fourths as much petroleum as they could, in theory, have handled. Most refineries were underutilized, and as a consequence profit margins were slim as different operators competed for a limited amount of oil. Throughout much of the 1980s and 1990s, industry profits averaged 5 percent or less. Major oil companies, which produced, refined, and marketed their own petroleum products, were sometimes content to operate their refineries at a slight loss because they made their profits in other segments of the industry.

Business strategies changed beginning in the late 1980s. Many refining companies merged or were bought out—others closed. Between 1995 and 2001 refinery closures resulted in the loss of capacity of approximately 830,000 barrels per day. By 2002, only 58 companies re-

the atoms in the hydrocarbon molecules together are broken as are the bonds that created the oxygen molecules. The carbon, hydrogen, and oxygen atoms now recombine to form carbon dioxide and water mained in the business, and the 10 largest firms controlled 80 percent of U.S. refining capacity. (Only five years earlier, in 1997, the 10 largest firms controlled 50 percent of U.S. refining capacity.) The situation remains essentially unchanged at present. The industry has responded to increased demand by improving existing facilities and increasing utilization rates, defined as the percentage of time that a facility is operating at its maximum capacity. Today, refineries operate at an average of more than 90 percent of their maximum capacity—that is, most refineries are working near their maximum output most of the time. These strategies have reduced competition and, perhaps predictably, profits have steadily increased. But in the refinery business, additional profits have not attracted additional competitors. Refineries are expensive to build and notoriously difficult to site. No large new refineries have been built in the United States since the 1970s, and recent attempts by the federal government to spur investment in new refineries have been thwarted on two fronts: (1) opposition based on the impact that a proposed refinery would have on the environment, and (2) the federal government's own policies encourage the use of nonpetroleum transportation fuels—a strategy that, if successful, would diminish the long-term need for additional refineries.

As the number of competitors has decreased, industry attitudes have changed as well. According to the RAND study, in the early 1980s, refiners generally accepted the importance, even the necessity, of maintaining a reliable supply of product, but as consolidation continued, a new idea took hold: If there were occasional, brief shortfalls in production, that was acceptable. In fact, occasional shortfalls lead to higher prices and potentially higher profits. Experience has already demonstrated that excess capacity leads to neither.

and trace amounts of other substances. The energy required to bind together the constituent atoms in the CO_2 and the H_2O molecules is less than the energy required to bind together the hydrocarbon and oxygen molecules present at the start of the reaction. The remaining energy appears in the form of heat. (Recall that energy cannot be created or destroyed, so the energy present at the start of the reaction equals the energy present at the end of the reaction.) The production of this heat is the purpose of the combustion reaction. The CO_2 and H_2O molecules produced by the reaction are essentially byproducts; they have no economic value and, in that sense, are incidental.

It is important to emphasize that a combustion reaction that produces only water and carbon dioxide is, from an environmental viewpoint, ideal in the sense that all other possible outcomes are, from an environmental point of view, worse. For example, incomplete combustion may produce carbon monoxide (CO), a poisonous gas; or if there are too many hydrocarbon molecules relative to oxygen molecules in the combustion chamber at the start of the reaction, not all of the fuel can be burned. In an automobile engine this means that the exhaust gases will contain uncombusted hydrocarbons as well as CO_2, H_2O, and CO. Antipollution equipment can complete the intended reactions by combusting the remaining hydrocarbons and CO to produce additional CO_2 and H_2O, but the antipollution devices act on the product gases *after* they have left the combustion chamber and, as a consequence, they contribute nothing to automobile efficiency. Carbon dioxide, of course, contributes to global climate change.

This is a fundamental problem associated with burning gasoline: Although it is in many ways the best transportation fuel currently available, the most efficient use of it entails the production of a greenhouse gas. There is no way to avoid production of CO_2, and there is currently no way to avoid venting it to the atmosphere. (The same general statements also apply to diesel and jet fuels.) Moreover, drastic reductions in the consumption of transportation fuels during the next few decades seems unlikely. In fact, the EIA estimates that consumption of petroleum in the United States will remain almost constant between 2009 and 2030. Rates of increase will be very high in India and China, where the automobile is only now on its way to becoming the standard means of personal transport.

There are other practical difficulties with using gasoline. One important problem occurs when the fuel-air mixture reacts too soon. The result is engine knock. Engine knock is noisy; it creates unnecessary wear on the engine, and it wastes gasoline.

Decades ago, automobile companies controlled the problem of engine knock by adding lead. As an additive for preventing engine knock, lead performed well, but lead is also a toxin and tests subsequently demonstrated that it was harming the health of many individuals, especially children. Beginning in 1979, a new antiknock additive was added to gasoline in place of lead called methyl tertiary-butyl ether (MTBE). Concentrations of MTBE in gasoline increased after passage of the 1990 Clean Air Act Amendments. As a so-called oxygenate, MTBE reduced the formation of smog and toxins.

Despite its value as an antiknock compound and as an oxygenate, ideas about the value of MTBE have changed just as they have changed for lead. Enormous amounts of gasoline are stored both above ground and underground, and occasionally there are leaks in the storage containers. The containers can be fairly large, as, for example, the underground storage tanks at service stations, or they can be quite small, as are gas tanks for pleasure boats. Occasionally, a tank leaks, and when a leak occurs the contents of the tank may become mixed with ground water. MTBE, because it is highly soluble in water, mixes with water more quickly than the other hydrocarbons in gasoline. Slow-moving spills of gasoline are, therefore, often fast-moving spills of MTBE. Tests indicate that MTBE can cause cancer in laboratory animals at high levels of exposure. It is not clear just how dangerous to humans MTBE is, either in terms of its health effects or with respect to the level to which an individual must be exposed before health effects appear. It is known that MTBE tastes and smells bad, so even if it were harmless from a public health point of view (and it is not), a gasoline spill could still make the affected water supply unpalatable. Many local and state governments have expressed concerns about the safety of MTBE as a fuel additive, and it has been gradually phased out to be replaced by ethanol. But in the United States, ethanol is produced mainly from corn, and it has become increasingly apparent that the large-scale production of corn-based ethanol is also associated with certain deleterious economic and environmental consequences. As farmers have grown more corn to meet the demand for ethanol, food costs have increased as agricultural resources have shifted away from the production of food and animal feed in favor of the production of fuel. And the manufacture of ethanol requires large amounts of water and energy as well as large government subsidies. All of this serves to demonstrate that the production of gasoline is now and has always been problematic from an environmental point of view.

Modern life is dependent on the consumption of enormous amounts of transportation fuels. Humankind's experience with acid rain, global warming, smog, and numerous other environmental effects associated with the consumption of gasoline, diesel, and jet fuel point to just how difficult it is to consume these products without wreaking widespread environmental damage. It has, in fact, long been apparent that the use of these fuels *must* entail widespread environmental damage. This knowledge has made little difference in consumption patterns, however, because relatively inexpensive transportation is valued so highly by so many. Oil remains the one primary energy source capable of powering a modern transportation system.

Patterns of Consumption

Every developed nation has become dependent upon a worldwide transportation sector to distribute even the most basic of goods. Air travel is commonplace, and commercial airlines are major consumers of petroleum. Buses and trains, which are fairly efficient users of petroleum, are another important market for transportation fuels, but the growth of low-density suburbs in some developed nations has compromised the ability of mass transportation to provide efficient service and led to a so-called "car culture," the creation of which presupposed ready access to large supplies of inexpensive gasoline. The first section of this chapter discusses petroleum's use as a transportation fuel with special reference to the United States, the world's biggest market.

There are, of course, other uses for petroleum than as a transportation fuel—heating oil, for example, continues to be burned in

The open road. Much of what Americans take for granted requires easy access to inexpensive petroleum. *(Scott Wilson)*

many nations. Describing some of these uses is the second goal of this chapter.

THE TRANSPORTATION SECTOR

Transportation in the United States is nearly synonymous with petroleum usage. Broadly speaking there are three main sectors to the transportation fuels market: gasoline, diesel, and aviation fuels. Each sector will be considered separately.

Gasoline is the most commonly used fuel in the United States. Approximately half of all oil used in the United States transportation sector is gasoline, and it is sold at 170,000 retail outlets. Average U.S. daily consumption of gasoline has exceeded 370 million gallons (1.4 billion l) per day since 2002, enough fuel for the nation's approximately 230 million cars and light trucks. In 2008, consumption averaged 377 million gallons (1.43 billion liters) per day.

The term *gasoline* refers to a variety of formulations, each with its own characteristics. There are, of course, the usual grades of gasoline—regular, mid-grade, and premium—and these are found throughout the nation. Less generally appreciated are the region-specific gasoline formulations.

In order to achieve various environmental objectives, retailers in different regions of the country are sometimes required to sell only gasoline that has been manufactured with certain additional characteristics. Emissions that result from burning these specially formulated gasolines may contain reduced levels of the poisonous gas carbon monoxide, or their combustion may contribute less to the formation of smog, or they may evaporate more slowly (thereby reducing air pollution), or they may have any of a variety of other characteristics that reflect the concerns of a specific region. What is important to keep in mind is that the region's retailers generally have no choice but to offer these formulations exclusively. This has made refining gasoline somewhat more complicated since it is no longer true that the old regular, mid-, and premium grades are suitable for all markets.

An unintended consequence of the trend toward regional fuel requirements is that the nation's gasoline supply system has become less stable. As mentioned in chapter 11, refineries now routinely operate at more than 90 percent of their maximum capacity, and they do not maintain large supplies of refined product on-site. This means that there is very little ability among refiners to compensate for a sudden drop in production due to an outage, anticipated or not, at the large refineries. Moreover, what little additional capacity exists in one region of the country is often useless to those in another region, because as a matter of law fuel from one region often cannot be marketed in another region if the formulations are different. While the federal government has begun to create regulations that seek to reduce the number of regional formulations—and so create larger internal markets—it is still true that gasoline supplies in the United States depend on almost everything working right all of the time, and when something goes wrong, fluctuations in supply and price occur almost immediately.

For the next decade, the gasoline supply situation in the United States can only change slowly. Although recent gasoline price increases have heightened public interest in more fuel-efficient vehicles, the EIA still predicts an increase in the number of cars and light trucks on the road over the next several years, thereby blunting the effect of more fuel-efficient vehicles on the fuel supply. Ethanol volumes will remain tiny in comparison to gasoline volumes, and refinery capacity will not change much because few, if any, new refineries will be built. Life in the United States depends on oil, and consumers are constrained in their response to price hikes by the lack of available alternatives. Gasoline prices increased 32 percent between the first quarter of 2007 and the first quarter of 2008, but according to the EIA, consumption declined only 0.6 percent. (This occurred during a period of significant inflation, and a growth in personal income of about 0.3 percent.) From a consumer's point of view, the fuel situation is not favorable, and there is little hope for improvement over the short term. No doubt an innovation in automobile design—an economical plug-in or battery-powered car, for example—would be warmly received by many consumers.

Diesel fuel is the second most commonly used fuel in the U.S. transportation sector, and diesel fuel is part of a larger market in what is called distillate fuel oil. Large trucks (approximately 2 million), locomotives (approximately 22,000), and many, but by no means all, buses, single-unit trucks, farm vehicles and military vehicles form part of the market for this fuel, but there are also non-transport applications such as, for example, home heating oil. The amount of distillate fuel oil consumed in the United States in 2008 was approximately 166 million gallons (627 million liters) per day. About three-fourths of the distillate consumed nationwide was for transportation.

Perhaps the most important difference between diesel and gasoline is that diesel has a higher energy density. Diesel yields approximately 11 percent more thermal energy per unit volume than gasoline. Functionally, the gasoline-air mixture burned in an ordinary automobile engine is ignited by a spark, and the diesel-air mixture burned in a diesel engine is ignited by compression. Diesel's use has long been associated with higher levels of pollution than that caused by gasoline consumption—particularly with respect to sulfur emissions and particulates—but the difference between the two fuels has narrowed considerably.

There are two main trends in diesel consumption. The first is the move to ultralow-sulfur diesel. Low-sulfur diesel was formulated to have a sulfur content that did not exceed 500 ppm (parts per million), but what was once considered low is now considered high. New ultralow-sulfur diesel has a maximum sulfur content of 15 ppm. By December 1, 2010, all new diesel fuel sold in the United States for use on the highway must be ultralow sulfur. The other main trend affecting diesel usage is the introduction of an alternative fuel. Biodiesel, which is produced from plant matter—and in the United States soybeans are often used as the feedstock—is now often found in diesel fuel in the form of a biodiesel-diesel blend. The properties of biodiesel are similar to those of ultralow-sulfur diesel, except for the fact that biodiesel has better lubricity—that is, biodiesel lubricates certain engine components better than ultralow-sulfur diesel as it flows toward the cylinders. (Fuel pumps in diesel engines depend on the fuel that flows through them for lubrication, and ultralow-sulfur diesel, unlike low-sulfur diesel, the fuel that it replaced, has poor lubricity properties.) Two percent biodiesel is sufficient to restore the necessary lubricity to diesel fuel. Consequently, the move to ultralow-sulfur diesel has created a market for biodiesel. The market for conventional diesel fuel is large enough, however, that displacing a large percentage of diesel fuel with biodiesel is currently impossible, because, as with ethanol, producing sufficient quantities of biodiesel for it to become a major transportation fuel would disrupt the production of food and feed.

An interesting change in the pattern of diesel consumption is occurring in the nation's fleets of municipal and school buses, many of which have invested heavily in a variety of alternative fuels. Most of these fuels—compressed natural gas is one of the fastest growing alternatives—provide much less energy per unit volume than diesel

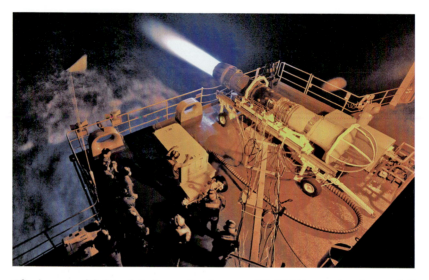

Afterburner of fighter jet. An engine from an F/A-18C Hornet is taken to full afterburner during a test by Aircraft Intermediate Maintenance Department's jet shop on the fantail of the USS *Harry S. Truman*. The Defense Department is a major consumer of transportation fuels. (*Petty Officer 3rd Class Kristopher Wilson, U.S. Navy*)

and are not available at ordinary retail outlets, but school buses and municipal buses operate on routes of fixed length and generally obtain fuel at special locations. Therefore the disadvantages associated with limited range and limited fueling sites do not apply to them. This sector of the transportation market is the one seeing some of the most significant changes in consumption patterns.

The third major segment of the transportation market is aviation fuel, almost all of which is jet fuel. In 2008, the United States consumed an average of 65 million gallons (245 million liters) of jet fuel per day. About 80 percent of the total amount of jet fuel was consumed by commercial aircraft. Military aircraft and some airport ground vehicles consumed the rest.

Jet fuel is burned under a wide variety of conditions. High-altitude flights take planes into cold regions of the atmosphere. The weather at the airport where the jet lands or takes off may be cold or warm. These factors, together with safety considerations, help explain the chemical composition of jet fuel. It ignites at a higher temperature than, for example, gasoline, and so is safer for use, but in extremely cold weather it is still able to flow. In fact, when the weather has become very cold, jet fuel has sometimes been blended with diesel and heating oil in order to prevent these fuels from thickening. (A thick or highly viscous fuel will not flow.)

Alternative fuels have had little impact on the jet fuel market so far, but the U.S. Department of Defense is encouraging the development of alternative fuels to ensure energy security, and researchers are testing fuels derived from natural gas, coal, and genetically engineered organisms. A B-52 aircraft has been successfully flown on a fuel derived from natural gas using a Fischer-Tropsch process (see chapter 6). Given the right technology, rapid changes in aircraft fuel are possible, but it has not happened yet.

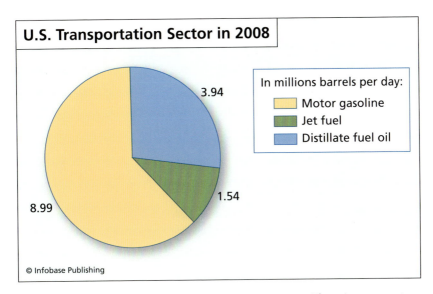

Petroleum consumption in the U.S. transportation sector. There is at present no substitute for any of these fuels, and each is critical to the national economy.

Alternatives to Gasoline

Gasoline has proven itself to be an excellent transportation fuel, but its production, storage, and use all have negative consequences with respect to the environment. Whether, overall, gasoline is a good value depends on the alternatives. The following is a list of some of the more probable ones.

1. *Natural gas.* The engine modifications required to run cars and trucks on natural gas are neither expensive nor difficult to install, and natural gas burns cleaner than gasoline or diesel. But there is no distribution network for natural gas as an automotive fuel. Natural gas can be used in one of two ways: compressed natural gas (CNG) or liquefied natural gas (LNG). Even when compressed to 204 atmospheres, a unit volume of compressed natural gas has only one-fourth as much energy as the same volume of gasoline. Consequently, CNG vehicles must have huge fuel tanks or short cruising ranges. LNG has a higher energy density than CNG, but it is more expensive. Maintaining LNG in a liquid state within an onboard automotive storage system presents a set of expensive and not-entirely-resolved safety challenges.

2. *Ethanol.* One benefit of ethanol is that it can be produced domestically. Another is that, sold as an ethanol-gasoline blend, it can be distributed using some of the same infrastructure used to distribute pure gasoline. On the negative side, ethanol requires a great deal of energy and water to produce. Moreover, the decision to produce fuel instead of food is already causing an increase in food prices, a trend that is expected to continue. Nor is ethanol as energy-rich as gasoline: Burning a unit volume of ethanol releases approximately two-thirds as much energy as the same volume of gasoline, so as the percentage of ethanol in an ethanol-gasoline blend increases, the consumer will drive fewer miles per tank of fuel. Finally, ethanol cannot be produced

in sufficient quantities to replace more than a small fraction of the gasoline supply.

3. *Hydrogen gas.* Touted as the fuel of the future, hydrogen is clean-burning and energy-rich. Currently, however, there is no infrastructure for distributing hydrogen, no infrastructure for producing large amounts of hydrogen at a cost that consumers can afford, and no method for safely and cost-effectively transporting hydrogen on board automobiles in sufficient quantities to enable consumers to travel more than short distances between fill-ups. Hydrogen may prove to be the fuel of the future, but it is a future that is at least decades away.

4. *Synthetic gasoline.* Shale, coal, and natural gas can be used to produce synthetic gasoline, but the processes are expensive and much harder on the environment than is the production of gasoline.

5. *Batteries.* Battery-powered automobiles have been around as long as gasoline-powered automobiles. They are clean and quiet, but limited in range and slow to charge. However, battery technology is improving. The great advantage to batteries is that they can be charged from the grid. Unlike hydrogen, for example, it is not necessary to build a new distribution infrastructure. Over the short-term, battery-powered cars may prove to be the consumer's best hope for reducing gasoline dependence.

Gasoline will be difficult to replace. Despite high prices and the environmental consequences caused by reliance on petroleum, the EIA's 2008 Annual Energy Outlook predicts a modest but steady rise in the consumption of transportation fuels, even after taking into account higher "corporate average fuel economy (CAFE) standards for light duty vehicles ... slower economic growth ... (and) the impact of higher fuel prices."

OTHER USES OF OIL

In the United States and many developed nations, less than 30 percent of all crude is used for applications other than transportation. From the refiners' point of view, these are secondary sources of profit, but from society's point of view, some of these applications are as essential to modern life as gasoline, diesel, and jet fuel.

A secondary source of profit for refiners—it is essentially a by-product of the refining business—is the home heating oil market. In the United States, about 7 percent of all households heat with residential heating oil. Most of these are located in the Northeast. Many homes originally built to use fuel oil have since switched to natural gas, but there remain about 8 million households that rely upon heating oil, which, as mentioned previously, is chemically very similar to diesel fuel. Refineries produce heating oil all year long, but it is only used in quantity during the winter months. A cold winter will draw down stocks of heating oil, but refineries do not show great flexibility in responding to temporary spikes in demand, because in order to produce large amounts of additional heating oil, they would have to refine large amounts of petroleum and in the process produce other products for which the demand has already been met. This would drive down prices and reduce profits. From the refiner's point of view, therefore, ramping up production of residential heating oil is a money-losing proposition. As a consequence, the supply (and so the price) of heating oil is unstable and subject to short-term price spikes. Not surprisingly, the spikes occur during the winter when the consumer is most vulnerable.

Gasoline, diesel, heating oil, and jet fuel constitute almost three-fourths of the yield from an average barrel of oil. After these fuels have been produced, what remains is ultimately converted into a larger number of products. Lubricating oils and greases, paraffin wax, asphalt, and bunker oil (used to power ships) are examples of products derived from the remaining petroleum.

Also extracted from non-fuel portion are so-called feedstocks for the petrochemical industry. Feedstocks are the raw materials from which many other, more familiar materials are derived, including polyester fibers, nylon fibers, acrylic fibers, epoxy glue, plastics of all sorts, ethylene glycol (a key ingredient in certain types of antifreeze) solvents, and hundreds of other familiar products. As with the materials described in the previous paragraph, the production of feedstocks for the petrochemical industry is a relatively small part of the business of refining, but the transformation of these feedstocks into useful materials is a key industry. Indeed, the petrochemical industry makes modern life possible.

In the end, what remains after the refining process is complete is a hard coal-like material called petroleum coke. It is essentially carbon, but it also contains sulfur and whatever trace metals were present in the original petroleum. Its exact composition depends upon the details of the process used to produce it as well as the composition of the petroleum from which it was derived.

Petroleum coke is, in a sense, the dregs of the refining process. There is little profit in its sale. Some is used domestically, but much of it is exported for use in power generation or as a heat source in kilns. With the disposition of petroleum coke, the petroleum is used in its entirety.

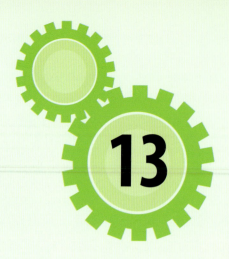

Environmental Costs of the Oil Economy

Oil has long been indispensable as a transportation fuel. No other energy source has proven itself so flexible, so energy-rich, and so convenient to use in the transportation sector as oil. But from an environmental prospective, the production and consumption of oil have been disastrous. Whatever one's feelings are about the world's oil dependence and its effects on the environment, most people indicate by their actions that they are willing to trade additional environmental degradation for rapid, reasonably priced, petroleum-fueled transportation. This was true a century ago, when the automotive and aircraft industries were in their infancy and their environmental impact was small, and it remains true today, when some of the world's largest industries are automobile manufacturers, aircraft manufacturers, and oil companies. The purpose of this chapter is to acknowledge some of the effects that the world's oil economy has had on the environment.

Exxon Valdez grounded on Bligh Reef and spilling oil, March 26, 1989 *(E. R. Gundlach)*

LOCAL EFFECTS

Local effects are those that affect the environment in the immediate neighborhood of the activity in question. Many of the environmental issues associated with the production and consumption of oil are local effects.

Local effects associated with oil exploration, activities that include the generation and collection of seismic data and the collection of core samples, are often not controversial. But occasionally assertions are made that these activities also are associated with substantial ecological costs. (See, for example, the sidebar "The Arctic National Wildlife Refuge.") Once oil is recovered, however, it must be transported, an activity for which the environmental effects are less ambiguous.

Oil tanker accidents have generally received the most attention in the popular press. Although tanker accidents, large and small, continue to occur, a great deal of improvement has been made within the industry both in terms of limiting the frequency of tanker spills and in terms of reducing their size. According to the International Tanker Owners Pollution Federation, which maintains a very complete set of statistics on all of the oil spilled from tankers in accidents in which more than 50 barrels were spilled, the trend is clear: 23 million barrels were spilled during the 1970s in 788 accidents, 8.6 million barrels were spilled during the 1980s in 449 accidents, 8.4 million barrels were spilled during the 1990s in 348 accidents, and while data for the first decade of the 21st century is incomplete, its total will almost assuredly be much lower than the amount spilled during the 1990s. From the beginning of 2000 until the end of 2008, only 1.7 million barrels of oil were spilled in 150 accidents in which the quantity of oil spilled exceeded 50 barrels.

The environmental impact of each large tanker accident is unique. To be sure, large slicks always have the potential to devastate local ecosystems, but that is not always what happens. The effects of a tanker spill depend on whether the wind is blowing towards shore or away and whether the turbulence in the sea impedes the formation of a large slick. It depends on the amount of time elapsed before emergency crews begin to work, the type of oil spilled, the temperature of the water, and many other factors. The effects are neither mysterious nor hard to identify. Crude oil tends to be sticky and to cover small organisms, thereby preventing them from breathing, and once covered with oil larger animals, such as sea otters or birds, animals that depend upon their fur or feathers to keep warm, may die from the cold because the oil compromises their ability to retain body heat.

Exposed to the sun, the potential of a spill to create environmental havoc will diminish rapidly. The lighter fractions of oil will evaporate, and what remains behind will, over a period of weeks, solidify. The consequences of spilled refined petroleum are very different from those of crude oil. Gasoline and diesel evaporate quickly, but they also tend to be more immediately toxic to

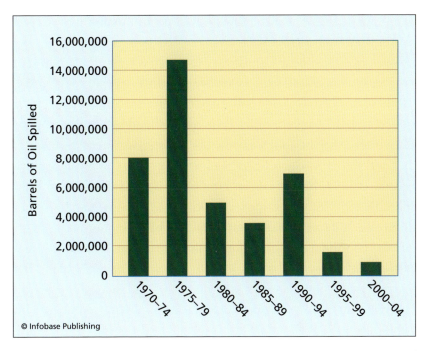

© Infobase Publishing

Tremendous progress has been made in limiting both the number and size of tanker spills. *(The International Tanker Owners Pollution Federation Limited)*

organisms exposed to them. Most large spills involve unrefined petroleum.

Tanker accidents are better known than most other types of spills because they are more dramatic. But in an average year, more oil leaks into the ocean through natural seeps, oil runoff from roads, and other small diffuse sources than from all the oil tanker spills occurring that year. The effects of these many smaller leaks are harder to quantify, however, because the composition of each small spill is unique, its individual effect is small, and the characteristics of the area affected by the spill vary greatly.

Pipelines are no less likely to pollute than oil tankers, even though pipeline spills seldom receive as much media attention as tanker spills. On October 1, 1994, for example, far more oil was lost in a pipeline break near the town of Usinsk in Russia than was lost during the 1989 *Exxon Valdez* tanker accident. (Estimates on the amount of oil lost near Usinsk vary from a few times the amount of the *Exxon Valdez* spill to as much as eight times its volume.) During the 1994 accident, oil poured across the tundra and into rivers. It spread across 72 square miles (190 km²) of land before freezing for the winter. And the spill was only the last step in an environmental disaster. Actually, the oil had been leaking from the pipeline since February 1994, but had been contained behind an earthen dam specially built for the purpose. (In Russia at the time the construction of such dams was common practice.) In October heavy rains and the continual accumulation of oil overwhelmed the dam. What distinguishes this spill is that it was a very large *single* incident, but Russia's vast pipeline system is in disrepair, and each year more oil leaks routinely from smaller pipeline leaks—leaks that attract no public attention at all—than is lost in any headline-making big spill, the spill at Usinsk included. Routine oil leaks have saturated the soil in large areas of western Siberia and Chechnya. Nor are pipeline leaks a uniquely Russian problem. In March 2006, approximately 6,300 barrels of oil spilled across the Alaskan landscape from a leak in the 800-mile (1,300-km) Trans-Alaska Pipeline System. It was a small leak, but it went undetected for five days.

In poor nations, pipeline spills tend to be more common and less publicized than they are in wealthier nations, but all nations experience them. Pipelines are long, complex engineering projects subject to everything from corrosion to sabotage to earthquakes. The complexity of large pipelines, the vast distances that they cross, and the huge quantities of oil that they sometimes carry all mean that it is not possible to eliminate spills entirely. As with so much else in the oil industry, occasional oil spills are something that oil consumers tolerate in order to be assured of a steady supply of fuel.

Although spills are usually unplanned, that is not always the case. The largest oil spill in history was a deliberate act: It occurred early in 1991 when the Iraqi army began to destroy oil wells and oil storage facilities in Kuwait. Approximately 9 million barrels of oil were spilled, 8 million of which drained into the Persian (Arabian) Gulf. The resulting oil slick covered 600 square miles (1,600 km^2).

The next step in the journey from oil field to consumer is the refinery. Refineries release large quantities of pollutants into the air. To place refinery emission rates in perspective, it is important to bear the following three points in mind. First, many refineries operate on a very large scale. (In the United States, upwards of 17 million barrels per day are processed.) Consequently, even when minute percentages of product are released into the environment—and this is exactly what happens in an efficiently operated refinery—the amount, when measured in pounds (or kilograms), will be large. These large numbers reflect the scale at which the industry operates.

Second, there is substantial uncertainty about the scale of refinery emissions. The major reason for the uncertainty involves so-called fugitive emissions, which emanate from leaking valves, storage tanks, and other pieces of refinery equipment. Refineries attempt to account for fugitive emissions when reporting total emissions, but there is disagreement about the reliability of their figures. Most sources of fugitive emissions are not large. A leaking valve, for example, might be a source of fugitive emissions, and a single valve does not constitute a large source. But each of the very large refineries that produce most of the nation's supply of petroleum products has many thousands of valves. A small percentage of valves left partially open will translate into a high emissions rate. In an important study of fugitive emissions prepared in 1999 for U.S. Representative Henry A. Waxman, Democrat of California's 30th congressional district, it was estimated that unreported fugitive emissions from U.S. refineries accounted for the majority of emissions from these facilities. While much may have changed since Congressman Waxman's study was published, the facts about the magnitude of fugitive emissions are no clearer nor are they less in dispute. In the words of Congressman Waxman's study, they remain, ". . . one of the single largest sources of pollution in the United States."

Third, a number of refineries are located near the major markets that they serve. Questions about the type and amount of pollution emanating from these industrial sites are of immediate importance to many.

But answers to these questions are not so obvious. Consider, by way of example, benzene. Refineries release large quantities of benzene into the atmosphere—at least hundreds of thousands of pounds per year in the United States alone. (There is substantial variation in the available estimates.) Some benzene is unavoidably released in the normal course of refinery operation, and some is released as fugitive emissions. But more relevant than the total benzene emissions of the refinery industry is the actual exposure of individuals to this hazardous chemical. Although benzene is harmful and easily absorbed, once benzene has been released into the air, it breaks down over the course of a few days. Nor does it accumulate in the body. In addition, everyone is regularly exposed to some benzene, and much of the exposure is voluntary. Paint, glue, the air around gas stations, and even cigarette smoke are significant sources of benzene. Here is what research has revealed: Prolonged exposure to elevated levels of benzene can damage one's immune system; benzene causes certain types of leukemia, especially acute myelocytic leukemia (AML); and refinery workers and those involved in the shoemaking industry have elevated levels of AML. The effect of benzene emissions from refineries on the health of those living near these important manufacturing concerns is less clear. The fact that refineries continue to operate—and often near major population centers—is one more indication of the value people place on plentiful and affordable transportation fuels.

Once the principal transportation fuels, all of which are liquid, are manufactured, they must be stored. Leakage from storage containers involves another set of hazards.

Finally, the fuel is burned. As described in chapter 11, when gasoline is burned under ideal conditions, the products of combustion are water and carbon dioxide. Under more realistic conditions, toxic carbon monoxide, nitrous oxides, which contribute to acid rain, and unburned hydrocarbons, which contribute to smog, are emitted along with water and carbon dioxide. All modern automobiles are equipped with various emission control technologies, and these have been instrumental in greatly reducing the amount of pollution emitted per car. But as the amount of pollution emitted per vehicle has plummeted, the number of vehicles on the road has soared. Consequently, the total amount of pollution in the area has declined more modestly than "per vehicle" figures indicate. Similar statements hold for diesel and aviation fuel.

GLOBAL EFFECTS

The one truly global effect of burning fossil fuels of all types arises from the production of carbon dioxide (CO_2). As described in chapter 4, CO_2 is a greenhouse gas. As the amount of CO_2 in Earth's atmosphere

As global temperatures increase and sea levels continue to rise, some low-lying islands will disappear beneath the waves. *(Wikipedia)*

The Arctic National Wildlife Refuge

The Arctic National Wildlife Refuge (ANWR) occupies 19 million acres (7.7 million hectares) of the 48.8-million-acre (19.8-million-hectare) North Slope region of Alaska. Substantial deposits of oil are present on the North Slope in general and inside ANWR in particular. This has been recognized for a long time. Exploration of the area began in earnest during the 1940s, and the North Slope, outside of ANWR, has already produced more than 15 billion barrels of oil. Many billions of additional barrels are technically recoverable. With so much experience, a great deal is now known about the effects of drilling in this environment. This information has fueled debate as public attention has turned to the possibility of drilling in the wildlife refuge itself—specifically, the debate is whether or not to drill inside a 1.5-million-acre (610,000-hectare) parcel called area 1002, an area consisting largely of tundra.

Tundra is flat, treeless land covered with lichens, mosses, and dwarf shrubs, and beneath the thin layer of soil upon which these plants depend, the ground is permanently frozen. Tundra is fragile in the sense that it recovers only slowly from surface damage. The fragility of the tundra is illustrated by the effects of earlier seismic studies conducted on North Slope tundra outside of ANWR. To perform the surveys, oil companies drove large trucks back and forth across the tundra. Decades later, the tracks left by these vehicles are still visible. Later, to extract the oil discovered during the surveys, companies built roads across the tundra so that they could move equipment from site to site, and when they drilled, they placed their equipment on large pads, beside which they placed "reserve pits," places for storing drilling waste. All of these activities created essentially permanent changes to the tundra. The proof of this statement

increases, more of the Sun's energy is retained by the atmosphere, and the average temperature of Earth's atmosphere and oceans increases. In the United States, the two biggest sources of CO_2 are power plants and the transportation sector—mainly cars, trucks, and airplanes. At lies in the fact that the locations of earlier and long-abandoned sites are still clearly visible.

But technology has improved. Today, reserve pits are no longer used because drilling waste is injected deep beneath the surface. The size of the pads has been reduced by 80 percent over those constructed in the 1970s, and roads are now built of ice and snow and only used during the winter, thereby protecting the tundra. When the summer arrives the roads melt, leaving no trace. Fewer pads are required, because companies can drill horizontally so that one relatively small pad can be used to access a larger area. Moreover, less drilling is required because companies strike far fewer dry holes due to improvements in exploration technology. Are these improvements enough to protect the environment and justify drilling? Those in favor of drilling say that the situation is much better than it was, and drilling can now be done responsibly. Those opposed to drilling emphasize the impossibility of producing oil without changing the landscape as well as the possibility of accidents that might make the situation worse.

The two sides often interpret the same information differently. For example, those in favor of drilling in area 1002 emphasize the small amount of land actually required for drilling activities—small, that is, compared to earlier drilling sites. Those opposed emphasize that the cumulative effects of multiple drill sites spread over a large area will be out of all proportion to the small pad size of individual drill sites.

This dispute shows how difficult it has become to supply gasoline, diesel, and jet fuel to those who demand it, and it also demonstrates that there is almost no place left on the planet that oil companies are unable and unwilling to drill.

present, essentially all of the CO_2 produced in the power generation and transportation sectors is vented directly into the atmosphere.

A key difference between the transportation sector and the power-generation sector is that only the power-production sector can substantially curtail CO_2 emissions over the next 15 years. Stationary sources of CO_2 can, in principle, be outfitted with devices to capture and store CO_2. There are, as mentioned in the section on coal, still substantial problems to overcome before sequestration can be implemented throughout the power-generation sector, but reasonable-sounding proposals for implementing sequestration technology in the power-generation sector are already under consideration. This is not the case in the transportation sector, where sequestration technology would be much harder and more costly to implement. A further distinction between the two sectors is that there are alternatives to fossil fuels in the power-generation sector. Indeed, some countries that were formerly dependent on fossil fuels to generate electricity, most notably France, have already learned to generate virtually all of their electricity without them. There are currently no large-scale alternatives to petroleum. It is indispensable in the transportation sector.

Supposing that an attempt were to be made to introduce sequestration technology into the transportation sector. What would be the barriers?

First, it is important to emphasize that every barrier that exists to the introduction of sequestration in the power-generation sector also exists for its introduction to the transportation sector. In particular, there is currently no infrastructure for sequestering large amounts of CO_2. But the problems associated with sequestration are more acute in the transportation sector. To sequester the CO_2 generated by burning gasoline, for example, hundreds of millions of automobiles would have to be outfitted with devices to capture and store the CO_2 on board. This would be a prohibitively expensive proposition even if the onboard technology were immediately available, which it is not. And numerous neighborhood collection stations would have to be built to off-load and store the CO_2 from automobiles and trucks. In addition, a pipeline system would have to be constructed to transport the CO_2 to a main collection point for sequestration. Putting aside the considerable technical difficulties involved, the costs of such an undertaking would be prohibitive.

Are there any solutions to the ever increasing amounts of CO_2 emitted into the atmosphere by the transportation sector? In the short-term, the answer is no. Even if consumers switched to more fuel-efficient vehicles, which would mean that each new vehicle emitted less CO_2 on average than the vehicle it replaced, the rising number of vehicles in the United States and elsewhere would, according to many analysts, overwhelm the individual reductions that would occur as a result of increased efficiency. The rate at which CO_2 is added to the atmosphere would, under these conditions, be less than it would have been without the move to higher fuel efficiency, but any decrease in the current global rate of emissions would be modest at best.

Over the longer term the solution to CO_2 emissions in the transportation sector may simply depend on switching fuels. Battery power, perhaps, or hydrogen, or some other power source that is now classified as "alternative," may hold the key.

Oil Markets and Government Policies

The price of oil is highly volatile, and experts have had only a modest amount of success in predicting price fluctuations. But while the price of oil has sometimes changed rapidly and unpredictably, the factors affecting that price have not. The oil markets are a unique mixture of anticompetitive pricing and competitive market forces coupled with consumer demand for refined products that is, over the short term at least, fairly inflexible. Part of the purpose of this chapter is to describe some of the factors affecting the price of oil.

Peak oil production, that point in history when the rate at which oil is produced cannot be further increased, is a reality. But as pointed out in chapter 8, there is substantial uncertainty about when peak oil will occur. Currently, there is general agreement that it will occur in a few decades, but there was general agreement a few decades ago that peak oil would occur "in a few decades." So far

National energy policy has helped to shape the oil economy as it exists today. *(Architect of the Capitol)*

there is no sign that peak oil has occurred. Given these uncertainties and past errors, what policies can and should be undertaken to prepare for peak oil? What, if anything, should governments do to prepare for peak oil? These questions are addressed in the second section of this chapter.

PRICING PETROLEUM

A fundamental feature of the petroleum markets is that there is little immediate flexibility in demand because, in contrast to other primary fuels, petroleum products are used primarily in the transportation sector, where they are currently irreplaceable. Increases in the cost of gasoline, for example, have historically had only a modest effect on consumption, since for most people there is no immediate alternative to the automobile and no practical alternative to gasoline as a fuel. In the face of price spikes, therefore, most consumers will, of necessity, continue to consume a certain amount of gasoline no matter the cost. Insofar as the price of gasoline reflects the price of petroleum, therefore, the price of petroleum has so far had little effect on the demand for gasoline.

Faced with increasing prices, consumers are not entirely helpless, of course. Given time and sustained high prices, they will eventually decide to switch to more fuel-efficient vehicles. They may eventually move closer to work, shopping centers, or schools, or live where there is better mass transit. But so far housing patterns have not been affected by the price of petroleum, and the occasional trend toward more fuel-efficient vehicles, a change that has occurred in response to some price hikes in the past, has always taken place slowly and against a backdrop of steadily increasing numbers of automobiles. As a consequence, the impact of any trend toward the purchase of more fuel-efficient vehicles on total petroleum consumption has, so far, been somewhat masked by the ever-increasing number of automobiles on the road.

Historically, gasoline consumption, when averaged over several years, increases. In the United States, for example, consumption has risen almost continually since the end of World War II. In the 50-year period from 1945 to 2008, U.S. gasoline consumption increased each year with only ten exceptions. Consumption decreased from 1951 to 1952 and from 1973 to 1974; it decreased from 1978 to 1979 and each year thereafter until the end of 1982; it decreased each year from 1988 through 1991 and again in 2008. These yearly decreases were small as a percentage of total consumption—the famous 1973–74 decrease, for example, a consequence of the OPEC oil embargo and OPEC price increases, amounted to only 2 percent of total gasoline consumption—but as small as these decreases in consumption were as a percentage of the total, each decrease marked a period of substantial economic disruption within the United States. Over the long term, U.S. gasoline consumption increased from approximately 579 million barrels in 1945 to 3.3 billion barrels in 2008, and the EIA expects consumption to continue to increase. Huge supplies of petroleum continue to remain a precondition for vibrant national economies. Faced, therefore, with strong demand, tight supplies, and a supply chain that is prone to disruption, how do buyers and sellers decide what the "right" price of petroleum should be?

On the supply side, the price of oil is the result of a complex interplay among several factors. Perhaps the factor that comes first to the minds of most people is OPEC, the Organization of Petroleum Exporting Countries. (See chapter 1 for an overview of OPEC.) To be sure, OPEC is an important factor influencing the supply (and so the price) of oil. This organization operates in a noncompetitive way in order to maintain a predetermined price for oil. To understand how OPEC can accomplish this, it is important to bear in mind that world oil supplies are often tight; that is, there is little excess capacity to meet spikes in demand. This means that a modest shortfall in production can have large impact on price, and OPEC member states account for roughly 40 percent of world production. To manipulate price, OPEC member countries agree to restrict petroleum production. Maximum rates of production for individual OPEC members are determined at meetings. In return for limiting production, each member state assures itself of a certain minimum level of profit.

Limits on the production of oil by individual states with the goal of limiting total production introduces a natural tension between what is best for the group and what is best for any particular member. Historically, agreements to limit production have sometimes led to "cheating" on the part of some members. Cheating occurs when a member country exceeds the production limit to which it previously agreed; it sells additional oil at a price that can only be obtained because other member countries have kept their production within the previously negotiated limits. Economically speaking, a certain amount of cheating benefits the cheater, but if there is too much cheating then the supply of oil will exceed demand, and the price of oil will tumble. "Excess" cheating, therefore, causes all members to lose in the sense that their profit per barrel decreases—their total profit may decrease as well—even as they increase production in order to compensate for the declining price of oil. To be sure, OPEC has had success in influencing the price of oil, but early expectations on the part of OPEC nations that they could simply dictate a price for oil have proven unrealistic due in part to cheating.

In contrast to attempts by OPEC to carefully manipulate the supply of oil, the oil supply can also fluctuate in response to difficult-to-predict events. Two events of particular interest are the 1990 invasion of Kuwait by Iraq, and Hurricane Katrina, which struck the United States in 2005. With respect to the invasion of Kuwait, neither Kuwait nor any of the nations that later came to the defense of Kuwait worked hard to avert the attack because they did not anticipate that it would soon occur. Similarly, while most interested parties knew well before Hurricane Katrina struck that Gulf Coast oil supplies *could* be disrupted by a powerful hurricane, meteorologists could provide only a few days warning that Hurricane Katrina *would* strike the Gulf Coast and so disrupt supplies. (Hurricane Katrina temporarily disabled 25 percent of U.S. crude-oil production.)

To stabilize petroleum supplies—and so petroleum prices—in the face of unanticipated disruptions, many nations maintain emergency stocks of oil. The largest of these is the United States' Strategic Petroleum Reserve, an enormous stockpile of petroleum, purchased at taxpayer expense, and set aside for emergency use. Early in 1991, the United States released petroleum to help alleviate concerns about disruptions to oil supplies due to the war with Iraq. A total of 21 million barrels were released. In 2005, the United States released 11 million barrels from the Strategic Petroleum Reserve in response to disruptions caused by Hurricane Katrina. These releases are small in comparison to the capacity of the Reserve, which is currently set at 1 billion barrels. (The United States also maintains a 2-million barrel heating oil reserve called the Northeast Home Heating Oil Reserve to assure a reliable source of heating oil for consumers in the Northeast in the event of an emergency.)

As the preceding paragraphs indicate, both producing and consuming nations are heavily involved in attempts to increase the predictability of world oil markets. Producing nations are concerned with assuring predictability of price, and consuming nations are more interested in assuring predictability of supply.

Operating alongside governmental efforts to manage petroleum supplies and prices is a very vibrant commercial sector. But commercial institutions generally operate by a different set of objectives than governmental ones. This raises the problem of how to determine the economic value of a unit of crude oil that is acceptable to all parties.

There are three types of transactions by which crude oil is traded on world markets. The most common is a contractual arrangement between seller and buyer that states the volume of crude oil to be delivered over the duration of the contract and the price to be paid for the oil. The contract may, for example, cover a year and specify a schedule according to which various shipments of oil must be delivered. Most of this is common sense, but there is a problem with contracts, especially long-term ones—namely, how does one determine a fair price for oil? The problem becomes more apparent for a contract that specifies periodic deliveries. A good price for the first shipment is the prevailing price—but how is the prevailing price to be determined, either today or in the future? Determining the prevailing price is the problem of *price discovery*. One common method of price discovery is to examine the results of other oil transactions.

A second method of selling oil is via the spot market. Spot market transactions provide a method for balancing supply in the face of short-term fluctuations in demand. They fill a niche that long-term contracts cannot. A spot transaction is a single transaction—that is, it concerns a single shipment of oil—and the price is negotiated at the time the agreement is made. The transaction is completed shortly thereafter. Spot prices are good indicators of the "instantaneous" price of oil because neither the past nor the future price of oil is relevant to a spot transaction. When demand is high, the spot market price for a shipment of crude oil is high, and when demand is low, spot market prices are low. Spot market price information is collected and published by various trade publications because spot prices are considered good indicators of demand (and hence the value) for a particular type of crude oil at the time that the transaction was made. Records of spot transactions are used in the process of price discovery.

The third type of crude oil transaction involves contractual agreements to buy (or sell) oil at a future date but at a price agreed upon at the time the contract is signed. This is called a *forward contract,* and except for the time difference it is similar to a spot transaction. The buyer and seller will agree on the type of crude oil, the volume, the place of delivery, the price, and the date of delivery. There are several reasons that buyers and sellers might want to complete a forward transaction in the present rather than wait and complete a spot transaction in the future. The buyer might, for example, be motivated by the need to assure future supplies at the negotiated price; the seller might prefer the security of knowing that when the time arrives, there will be a buyer waiting to take possession of the oil at the negotiated price. Alternatively, the buyer might believe that when it comes time to complete the transaction the spot price of oil will be higher than the negotiated (preset) price, thereby enabling the buyer to "buy low and sell high"—the seller might believe that the spot price for oil will be less than the negotiated price at the time the transaction must be completed, in which case it makes sense to lock in the price early. Of course, when the time comes to settle, the spot price will almost certainly be either higher or lower than the price specified in the forward contract. If it is higher, then the buyer will have gotten the better deal; if the spot price is lower, the seller will have made the better deal. Because oil prices have a long history of fluctuating unpredictably, forward contracts enable buyer and seller to manage the risks involved in oil transactions in ways that are best for them.

Forward contracts, while they are useful, are custom-tailored for each transaction and so are not easily traded, a fact that makes it difficult for either party to sell its obligations to a third party should it want to do so. A more standardized version of a forward contract, called a *futures contract,* is designed to allow interested parties to more easily trade forward-like contracts. In the United States, oil futures are traded on NYMEX, the New York Mercantile Exchange. Futures contracts allow interested parties to account for the fluctuating price of oil on future profits and so distribute the risks associated with market fluctuations among many parties, some of whom have oil, some of whom want to take delivery of oil, and some of whom are only interested in the value of the contract. Contractually, the main difference between futures contracts and forward contracts is that future contracts are more "liquid"—that is to say, more easily traded—than are forward contracts. Futures contract are designed so that they are really "idealized" oil trans-actions. In fact, futures contracts seldom result in the delivery of any oil at all. They are, in effect, contracts about risk rather than oil. This has given rise to two distinct oil markets, the cash mar-ket, which involves delivery of oil, and the futures market, which involves delivery of oil contracts but seldom results in the delivery of actual oil.

Futures markets are designed so that the price of futures contracts moves up or down in parallel with the price of oil. (In the event that oil futures fail to move in parallel with the price of oil, the divergence between the cash and futures price can be quickly converted into cash by an alert futures trader.) The efficient functioning of the futures market does not require that the futures price equal the cash price, only that the futures price moves more or less in parallel with the cash price. People who buy and sell in the cash and futures markets are called traders. Traders may not own oil wells nor represent large consumers of oil. They may, and often do, operate more like grocers, who buy and sell food, but neither own farms nor consume the food that they sell. Some traders use futures contracts to reduce their exposure to risk in the cash market. This might work in the following (idealized) way: A trader may agree to sell x units of oil for y dollars and make delivery in one month. This is a cash transaction. The trader may not own any oil at the time of the transaction, preferring instead to buy oil later, nearer the time of delivery, and avoid storage fees. Of course, the trader does not know the future price of oil, but like everyone else, the trader knows that the price of oil is volatile. If the price of x units of oil later increases to $y + m$ dollars, the trader will be obligated to buy the oil at $y + m$ dollars in order to deliver it at a price of y dollars. The trader will, therefore, lose m dollars in the transaction.

The trader can reduce the risk of losing money on the transaction by entering into a futures contract at the same time that the trader signs the cash contract. The trader will purchase a futures contract for x units of oil at a price of z dollars, and the contract will specify delivery of the oil in one month, the same term as the cash contract. Now suppose again that in one month the price for x units of oil on the futures market has increased to $z + m$ dollars. The futures contract can be sold for $z + m$ dollars. Because the cash and futures markets move in parallel, the trader will earn a profit of m dollars from the futures contract. Now the trader settles the cash contract, thereby losing m dollars. The loss on the cash market and the profit in the futures market cancel each other, and the trader's profit or loss is determined solely by the terms of the initial contract. By using a futures contract in this way, the trader has protected the initial contract from future fluctuations in the price of oil. The situation works in essentially the same way if the price of oil were to decrease m dollars, in which case the fluctuating price of oil would cause the trader to earn m dollars on the cash contract and lose m dollars on the futures contract.

Futures Summary

A trader signs a contract to deliver x units of oil in 30 days for a price of y dollars. The trader plans to buy x units of oil just prior to the delivery date in order to avoid storage fees.

> Case #1: The price of x units of oil increases m dollars between the time the contract is signed and the delivery date of the oil.

	CASH MARKET	FUTURES MARKET
Initial actions	sell oil for y dollars	buy futures contract for z dollars
One month later	buy oil for delivery at $y + m$ dollars	sell futures contract for $z + m$ dollars
Results	loss of m dollars due to increase in the price of oil	profit of m dollars due to increase in the price of oil

> Case #2: The price of x units of oil decreases m dollars between the time the contract is signed and the delivery date of the oil.

	CASH MARKET	FUTURES MARKET
Initial actions	sell oil for y dollars	buy futures contract for z dollars
One month later	buy oil for delivery for $y - m$ dollars	sell futures contract for $z - m$ dollars
Results	profit of m dollars due to decrease in the price of oil	loss of m dollars due to decrease in the price of oil

This strategy of using the futures markets to protect against fluctuations in the price of oil is called hedging. It does not protect the trader from the consequences of making an inept deal with respect to the original transaction, but it does protect the trader from unpredictable fluctuations in the cost of oil over the term of the contract. One consequence of hedging is that the trader cannot benefit from fluctuations in the price of oil. In return for more predictability, the trader surrenders the possibility of additional profit. Energy futures markets attract a great deal of money and only some of it is from traders wishing to reduce their risks. Other traders believe that they can profit from short-term fluctuations in the price of oil. They are essentially betting that they can predict the future price of oil. Traders who use the futures market in this way are called speculators. Oil markets are rife with speculators.

Information about futures contracts and spot market prices enable those who buy and sell oil to develop a clearer idea about the price of oil and how the price of oil can be expected to fluctuate. This information is useful for computing the contract price for oil.

The actual details of oil trading are more complex than was indicated here. The complexity has evolved out of relatively simple ideas as buyers and sellers have sought to create a system that is

Collusion or Free Market?

When there is an interruption or potential interruption in the supply of oil, the spot price of oil often increases dramatically in anticipation of a potential shortage. Within a day or two the cost of gas at the neighborhood filling station goes up as well. This causes resentment on the part of many consumers, and it is not hard to see why. The gasoline in the filling station tank—the gasoline on which the retail price was just dramatically raised—is from a shipment that was delivered before the price of oil went up on the international oil market. The same can probably be said about the next few shipments of gasoline as well. It will take time for the more expensive oil to be transported to the refineries, be refined, and then for the resulting gasoline to be transported to the retail outlet. All of the gasoline on which the price was originally raised was made from the older, cheaper oil. This is a windfall for someone. Is the increase in price a sign of anticompetitive practices?

The price increase does not of itself prove the existence of anticompetitive practices. Here is why: Between the oil wells and the consumer's gas tank there is a certain amount of gasoline in storage. The owners of the stored gasoline are seeking to maximize their profit by selling the gasoline at the highest price that they can get. News reports cause them to believe that the price of unrefined oil is about to go up. When products manufactured from this more expensive oil are distributed to retailers, the price of the new product will reflect the new and higher price of the crude oil from which it was refined. This much is sure.

But with respect to the disposition of the oil they had on hand when the price increase was announced, they have one of two choices. One possibility is that they can hold onto the gasoline for a while and then sell it at the same price as the more expensive and yet-to-be-delivered product. Alternatively, they can sell gasoline immediately at the lower price. Immediate sale, however, means they will fail to maximize their profits. For someone trying to maximize profits, this is a simple decision to make. They hold onto their gasoline and wait for the price to increase.

(continues)

(continued)

If enough individuals with gasoline in storage hold onto their gasoline in order to sell it at the better future price—a price that they believe their product will soon achieve—a regional shortage develops. It can happen almost immediately. But this particular shortage is not due to a shortage of product. Rather, it is a shortage of economically available product. It is a shortage with a price. The price needed to eliminate the shortage is the price that the gasoline storage facility operator expects to earn from the more expensive gasoline, provided the operator waits a few weeks. As soon as prices climb to the anticipated cost of gasoline, the shortage disappears. No collusion is necessary.

The owners of the storage facilities are acting reasonably, as are the owners of the retail outlets. While parties may (or may not) collude or, at least, want to collude to set prices, it is not necessary to appeal to anti-competitive practices to explain these types of price fluctuations.

responsive to changing market conditions while distributing risk among the participants in a way that is acceptable to all parties. These complex procedures have drawn a great deal of criticism especially with regard to the role of speculators, who, it is sometimes claimed, drive up the price of oil to make huge profits at the expense of the consumer. To be sure, there is a great deal of speculation in the oil markets, and huge profits are made, but there has never been a golden age of fair play in the oil business.

Some critics advocate banning speculators from the market. But should speculators leave, they would take their money with them. Speculators increase the pool of buyers and sellers. Their presence increases the probability that when a trader (of any type and for whatever reason) wants to find a buyer or seller for a futures contract, one will be available. This is called liquidity, and without liquidity markets could not function efficiently.

The current system has evolved, in part, to correct certain deficiencies in past practices. Today's oil markets are far from perfect, but whatever their defects, oil markets have, at least, been remarkably successful at delivering the goods on time and at a generally affordable price.

NATIONAL POLICIES

If, when world oil production finally peaks, oil is still the essential transportation fuel that it is today, peak oil will mark the beginning of an economic disaster. What can be done to prepare for peak oil?

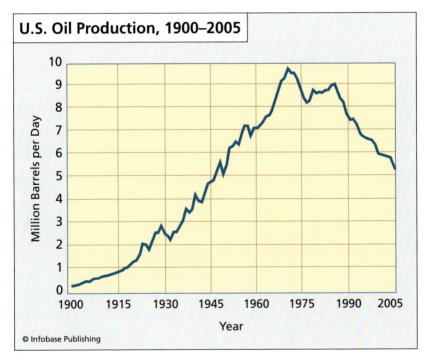

U.S. Oil Production, 1900–2005

© Infobase Publishing

Peak oil has already occurred in U.S. domestic production.

To see the way that consumption patterns might be affected by peak oil, consider the history of oil production in the United States. While the United States remains a major oil-producing nation, for many years it produced far more oil than any nation on Earth. During the first half of the 20th century it produced all of the oil required to meet its considerable domestic demands, and it exported oil all over the world. Oil production in the United States peaked in the 1970s, and as the chart on page 183 shows, production dropped precipitously thereafter. Of course, peak oil was not an immediate economic disaster for the United States because it could meet domestic shortfalls with ever-increasing amounts of imports, a strategy that it pursues to this day. Current models of a worldwide peak in production suggest a similarly steep drop in after-peak production, and, of course, this time imports are out of the question.

In 2007, the U.S. Government Accountability Office published "Crude Oil: Uncertainty about Future Oil Supply Makes It Important to Develop a Strategy for Addressing a Peak and Decline in Oil Production," a report that emphasized the uncertainty associated with the timing of peak oil production. Although peak oil is often described as a problem in physics, there are additional factors that complicate when peak oil may occur. The first such problem involves political instability. The report notes that roughly two-thirds of proven worldwide oil reserves are contained in countries with medium to high risks of political instability. While political instability does not change the amount of technically recoverable oil, it may change the rate at which it becomes available or the rate at which oil is consumed. (See the discussion of Iran later in this section.)

Second, although some countries freely provide the best available estimates about their proven reserves, some of the world's most important oil producers are not so open. The actual amount of oil in the Middle East, the world's major oil-producing region, is uncertain, in part because some of the most important oil producers

Emirates Palace in Abu Dhabi, site of 146th OPEC conference in 2007. Tremendous amounts of wealth are transferred from consuming to producing nations each day. *(Emirates Palace Hotel)*

from that region have not been forthright about providing realistic estimates of their reserves. The estimate for petroleum reserves in Kuwait, for example, did not change from 1991 until 2002, but during that time Kuwait produced 8 billion barrels of oil and made no new major discoveries of additional oil. Such a discrepancy calls into question estimates that depend in an essential way on knowledge of Kuwaiti reserves.

There are other questions about the ability of some OPEC members to remain net producers of petroleum that the report does not discuss. For example, gasoline in Iran has long been heavily subsidized by the government. Some contend that the Iranian government would fall if it did not continue to subsidize the price of gasoline. But heavy and prolonged subsidies have caused two trends: First, domestic gasoline consumption has increased far faster than new discoveries of petroleum, with the result that Iran has often failed to meet even the export quota that it has been allotted under OPEC agreements. Second, in the absence of even the potential for domestic profits, much of the Iranian oil infrastructure has fallen into disarray. As a consequence, Iran has become a gasoline importer. If these trends are projected into the future, there is the possibility that Iran will, despite its enormous petroleum reserves, become a minor exporter of oil, or even a net importer.

In the face of so much uncertainty, what can the United States—and by extension, other nations—do? First, the report asserts that additional attention needs to be paid to developing technologies that would delay the occurrence of peak oil—for example, the development of technologies that would enhance oil-recovery rates. Even now, using the best technologies available, more oil is often left in the ground than is recovered.

Second, according to the report, the government can invest more heavily in the development of alternatives to oil: Biofuels, coal, and oil shale are three possible (partial) replacements for oil. At present, these fuels have limited value either because they are too expensive to produce, which is the case with oil shale, or because they cannot be produced in sufficient quantities to greatly affect consumption patterns, which is the case with ethanol. If peak oil does not occur for several decades, there may be sufficient time for these fuels (or others) to displace significant amounts of oil. Given enough time and money, the problems currently associated with alternative fuels can, presumably, be solved. An emergency would arise, however, if peak oil were to occur sooner and with less warning.

Finally, bearing in mind that huge quantities of petroleum are needed only to supply the transportation sector, the report emphasizes the importance of further research into more fuel-efficient vehicles with an eye toward decreasing total demand. These vehicles might use petroleum more efficiently than today's cars, or they might not require petroleum at all, which would be the case if, for example, they operated on batteries or hydrogen. Developing vehicles that can effectively substitute for conventional automobiles, the report suggests, should receive more emphasis in government research programs.

This report has attracted a great deal of attention. It is concise and well-written, and it illuminates the technical difficulties future researchers face, but as with so many reports, it is painfully narrow. There is, for example, little attention given to the impact that some of these strategies will have on the environment. The development of shale oil and coal-to-liquids technology, in particular, will almost certainly entail the production of copious amounts of greenhouse gases and an unprecedented level of mining activity with all its attendant human and environmental costs.

For a century, oil has been irreplaceable as a transportation fuel. The consumption of oil in the form of gasoline, diesel, and jet fuel make modern life possible, and at the same time, society's oil dependence is responsible for substantial environmental, economic, and national security costs. Moving away from oil before world production peaks would be a historic change; moving away from oil after world production peaks would be a historic disaster. These are exciting times.

Chronology

1712 Thomas Newcomen builds his first steam engine

1765 James Watt invents the condenser, his most significant invention

1792 William Murdock uses coal gas to illuminate his house in a demonstration of the technology

1816 Baltimore becomes the first city in the United States to install gas street lights, using coal gas as an illuminant

1842 Miners' Association of Great Britain established

1859 First commercial oil well sunk at Titusville, Pennsylvania

1861 First meeting of the American Miners' Association

1869 110 miners killed in single accident at Avondale Mine in Plymouth, Pennsylvania

1870 Standard Oil formed by J. D. Rockefeller, M. B. Clark, and H. M. Flagler

1890 United Mine Workers of America founded in Columbus, Ohio

1891 109 miners killed at Mammouth Mine in Mount Pleasant, Pennsylvania

1897 United States coal production first reaches 200 million short tons (180 million metric tons)

1901 Discovery of Spindletop, the first major oil field in Texas

1902   United States coal production first reaches 300 million short tons (270 million metric tons)

1906   United States coal production first reaches 400 million short tons (360 million metric tons)

1907   Explosion at mine in Monongah, West Virginia, kills 362 miners

1911   Breakup of Standard Oil

1915   United States gasoline production overtakes kerosene production for the first time

1948   Massive Ghowar oil field discovered in Saudi Arabia

United States becomes net oil importer

1960   Formation of OPEC

1970   U.S. oil production peaks at 3,517,450,000 barrels per year

Passage of United States Clean Air Act

1973   Oil embargo initiated against the United States and the Netherlands

1975   U.S. Congress authorizes creation of the Strategic Petroleum Reserve

1979   Iranian Revolution ends as the Shah is forced into exile

1980   Arctic National Wildlife Refuge established

Construction of the Great Plains Synfuels Plant in Beulah, North Dakota

Start of Iran-Iraq War

1983   Crude oil futures contracts traded on the New York Mercantile Exchange for the first time

1986   World crude oil prices collapse

1989   The *Exxon Valdez,* a large oil tanker, runs aground, spilling its contents into Alaskan waters

1990   Iraq invades Kuwait.

U.S. coal production first reaches one billion short tons (900 million metric tons)

Clean Air Act renewed

1991   First Gulf War

Drawdown of stocks from U.S. Strategic Petroleum Reserve

1994   Massive pipeline spill near Usinsk, Russia

1996   The advanced Polk Power Station in Polk County, Florida, begins commercial operation; it uses integrated gasification combined cycle (IGCC) technology

1997   Kyoto Protocol, an international agreement aimed at reducing greenhouse gas emissions, is negotiated

2003   Second Gulf War

United States announces FutureGen project, a demonstration project that would produce electricity and hydrogen with very low emissions of any pollutants, including greenhouse gases

2005   Drawdown of stocks from Strategic Petroleum Reserve to compensate for disruption in supply due to Hurricane Katrina

2006   Sago mining accident; 12 miners die

2007   Crandall Canyon mine accident; nine miners die

2008   Oil reaches 120 dollars per barrel during the second quarter of the year, only to plummet to 60 dollars per barrel during the fourth quarter

United States shuts down FutureGen project just before the start of construction

# List of Acronyms

CCP	coal combustion products
CDC	Centers for Disease Control and Prevention
CWP	coal workers' pneumoconiosis (black lung disease)
EIA	Energy Information Administration
FGD	flue gas desulfurizer
IEA	International Energy Agency
IGCC	integrated gasification combined cycle
MTBE	methyl tertiary-butyl ether
OPEC	Organization of Petroleum Exporting Countries
UMWA	United Mine Workers of America

# Glossary

anthracite   a hard coal containing little volatile matter

ash   noncombustible materials found in coal

biodiesel   fuel formulated for diesel engines, made from vegetable oil or animal fats

biomass   plant or animal matter used as fuel

bituminous coal   a volatile-rich coal that is higher in rank than sub-bituminous and lower in rank than anthracite coal

coal gas   gas made from heating coal in a low-oxygen container

coalification   the process by which coal is made

cracking   a process in which heavier hydrocarbons are converted into lighter hydrocarbons

efficiency   the ratio of the energy delivered by a machine to the energy supplied to the machine

ethanol   colorless liquid derived from corn or other plant matter and increasingly used as automotive fuel

fluidized bed combustion   process for burning coal, sometimes with other fuels, in which the fuel is suspended in the combustion chamber by a stream of hot air

forward contract   a tailor-made contract between two parties that specifies the terms of a trade that will take place at some point in the future

futures contract   a standardized forward contract that can be traded on an exchange

gasifier   a device for converting coal to gas

global warming   the increase in the average temperature of Earth's atmosphere due to changes in the chemical composition of the atmosphere

greenhouse gas   a gas that if added to the atmosphere in sufficient quantities will cause global warming

heat engine   a device for converting thermal energy into work

heating value   a measure of the amount of heat released by a material during combustion

hydrocarbon   a molecule or material consisting of hydrogen and carbon atoms

lignite   a usually brownish-black coal of low heating value

longwall mining   a highly automated form of underground mining that extracts an entire panel of coal, leaving no pillars

metric ton   a unit of mass equal to 1,000 kilograms

oil sands   very viscous hydrocarbon (also called tar sands or bitumen)

price discovery   the determination, through market mechanisms, of the prevailing price of a commodity

products of combustion   any materials—gas, liquid, or solid—resulting from combustion

pulverized coal technology   a power-production technology designed to burn coal powder

reactants   substances, such as fuel and air, that enter into a chemical reaction

refinery   facility used to convert crude oil into fuel or other economically valuable products through a process based in part on distillation

roof fall   the collapse of the rock immediately overhead in a mine

room-and-pillar mining   the conventional method of coal mining, in which some of the coal is removed and the rest is left in the form of pillars to support the rock overhead

sequestration   any of a collection of technologies that isolates carbon dioxide from the broader environment for a prolonged time and in a controlled manner

short ton   unit of weight equal to 2,000 pounds

spot transaction   a contract for sale with delivery to take place immediately

subbituminous coal   coal with a rank higher than lignite and lower than bituminous

surface mining   any of several methods of extracting coal that involve scraping away the overburden and producing the coal from large open mines

turbine   a device used to convert the linear motion of the working fluid into rotary motion

volatile content   the percentage of matter in a sample of coal—exclusive of moisture—that is driven off by the application of heat

# Further Resources

Modern lifestyles are based on the consumption of large amounts of electricity—more often than not from coal-fired power plants—and oil, almost all of which is in the form of transportation fuels. Because these energy sources occupy so central a place in the lives of so many, they can be studied under many areas of knowledge, including history, engineering, science, public policy, or economics. The following is a list of some different ways of understanding these energy sources.

BOOKS

Clark, Stanley J. *The Oil Century: From the Drake Well to the Conservation Era.* Norman: University of Oklahoma Press, 1958. This book has been out of print for a while, but is still easily available through interlibrary loan. It is an important book because it is filled with numerous eyewitness accounts of some of the most significant events in the early history of the oil business. Highly recommended.

Errera, Steven, and Stewart L. Brown. *Fundamentals of Trading Energy Futures and Options.* 2nd ed. Tulsa, Okla.: PennWell Corporation, 2002. For those interested in the relationships that exist between physical and financial markets, this is a good introduction. The presentation involves no advanced mathematics, and the text is accompanied by a good glossary to help the reader decipher the financial jargon, which is both dense and unavoidable.

Goodell, Jeff. *Big Coal: The Dirty Secret Behind America's Energy Future.* Boston: Houghton Mifflin Company, 2006. A popular and passionate piece of investigative journalism.

Long, Priscilla. *Where the Sun Never Shines.* New York: Paragon House, 1989. A comprehensive look at the history of coal; it contains many interesting facts and stories.

National Academy of Engineering. *The Carbon Dioxide Dilemma: Promising Technologies and Policies.* Washington, D.C.: National Academies Press, 2003. This is a series of not-especially-technical papers by engineers, physicists, climatologists, economists, oceanographers, and others about global climate change and strategies to control it. This book can also be read on the web, albeit in a very user-unfriendly form. Available online. URL: http://books.nap.edu/openbook.php?record_id=10798page=R1. Accessed on July 10, 2008.

National Research Council and National Academy of Engineering. *The Hydrogen Economy: Opportunities, Costs, Barriers, and R&D Needs.* Washington, D.C.: National Academies Press, 2004. This book concentrates on describing what would have to occur in order to create an economy where hydrogen supplants gasoline. The most likely answer seems to be an increase in coal consumption, because coal would be the most likely feedstock from which hydrogen would be produced. This book can also be read on the Web in a very inconvenient form. URL: http://www.nap.edu/openbook.php?isbn=0309091632. Accessed on September 1, 2007.

Parra, Francisco. *Oil Politics: A Modern History of Petroleum.* London: I. B. Tauris & Company, 2004. This book covers the period from 1950 until 2000. It is interesting, detailed, and, for some reason, angry in tone. It is worth the read.

Peterson, D. J., and Sergei Mahnovski. *New Forces at Work in Refining: Industry Views of Critical Business and Operations Trends.* Arlington, Va.: RAND Corporation, 2003. A short, carefully written book, it provides analysis of changes in the refinery business and is based on interviews with individuals involved in the industry. The appendices provide the small amount of technical background necessary to follow the discussion on those rare occasions when it becomes technical. Highly recommended.

Pool, Robert. *Beyond Engineering: How Society Shapes Technology.* New York: Oxford University Press, 1997. An interesting book that is concerned with the general problem of how society affects technology. While not specifically about coal, oil, or power production, it offers general insight into the way that technology evolves.

Roy, Andrew. *A History of the Coal Miners of the United States, from the Development of the Mines to the Close of the Anthracite Strike of 1902, Including a Brief Sketch of Early British Miners.* Westport, Conn.: Greenwood Press, 1970. This book provides important information not easily available elsewhere about the early history of coal miners' unions. Originally published in 1905, it is now also available through the Cornell University Library Historical Monographs Collection. URL: www.library. cornell.edu. Accessed July 10, 2008.

INTERNET RESOURCES

The best source on the Web is the U.S. Energy Information Administration. Although its Web site is not easy to navigate, the articles are comprehensive and written in clear, easy-to-understand language.

Energy Information Administration. "A Brief History of U.S. Coal." Available online. URL: http://www.eia.doe.gov/cneaf/ coal/coal_ref_pdf/text.pdf. Accessed on September 1, 2007. A highly readable 43-page introductory-level report.
———. "Electricity Supply: Summary of the Kyoto Report." Available online. URL: http://www.eia.doe.gov/oiaf/kyoto/electricity.

html. Accessed on September 1, 2007. The Kyoto Protocol is an important document that has guided the energy policies of many nations, the United States and China being two notable exceptions.

————. "Oil Market Basics." Available online. URL: http://www. eia.doe.gov/pub/oil_gas/petroleum/analysis_publications/oil_ market_basics/full_contents.htm. Accessed on September 1, 2007. This is a book-length primer on oil, and it is well worth reading.

————. "A Primer on Gasoline Prices." Available online. URL: http://www.eia.doe.gov/bookshelf/brochures/gasolinepric-esprimer/eia1_2005primerM.html. Accessed on September 1, 2007. The price of gasoline is influenced by the price of oil, federal and state taxes, supply issues, environmental laws, and a number of other issues, some of which are fairly obscure. This brief article provides a nice introduction to the subject.

Government Accountability Office. "Crude Oil: Uncertainty about Future Oil Supply Makes It Important to Develop a Strategy for Addressing a Peak and Decline in Oil Production." Available online. URL: http://www.gao.gov/new.items/d07283. pdf. Accessed on April 30, 2008. A thoughtful report about peak oil.

International Energy Agency. "Control and Minimisation of Coal-Fired Power Plant Emissions: Zero Emissions Technologies for Fossil Fuels." Available online. URL: http://www.iea. org/textbase/papers/2003/Coal_Fired_Fossil_Fuels.pdf. Accessed on September 1, 2007. This is a nontechnical overview of coal combustion and ways that its environmental effects can be minimized.

Miller, C. Lowell. "Statement before the U.S. Senate Committee on Energy and Natural Resources, April 24, 2006." Available online. URL: http://fossil.energy.gov/news/testimony/ 2006/060424-C._Lowell_Miller_Testimony.html. Accessed on

September 1, 2007. This testimony by Mr. Miller, director of the Office of Sequestration, Hydrogen, and Clean Coal Fuels, Office of Fossil Energy, provides an overview of the challenges and advantages of producing synthetic petroleum from coal.

Fish and Wildlife Service. "Arctic National Wildlife Refuge: Potential Impacts of Proposed Oil and Gas Development on the Arctic Refuge's Coastal Plain: Historical Overview and Issues of Concern." Available online. URL: http://arctic.fws.gov/issues1. htm#section1. Accessed on September 1, 2007. Background on an important issue.

Index

Note: *Italic* page numbers indicate illustrations; page numbers followed by *c* indicate chronology entries; page numbers followed by *t* indicate charts, graphs, or tables.